Boiling Point

Also by Ross Gelbspan

The Heat Is On

Boiling Point

How Politicians,
Big Oil and Coal,
Journalists, and Activists
Are Fueling the Climate Crisis—
and What We Can Do to Avert Disaster

Ross Gelbspan

BASIC
BOOKS
A Member of the Perseus Books Group
New York

Hardcover first published in 2004 by Basic Books,
A Member of the Perseus Books Group
Paperback first published in 2005 by Basic Books

Books published by Basic Books are available at special discounts for bulk
purchases in the United States by corporations, institutions, and other organizations. For more information, please contact the Special Markets Department at the Perseus Books Group, 11 Cambridge Center, Cambridge, MA
02142, or call (617) 252-5298 or (800) 255-1514, or e-mail special.markets@
perseusbooks.com.

Library of Congress Cataloging-in-Publication Data
Gelbspan, Ross.
   Boiling point : how politicians, big oil and coal, journalists and activists
have fueled the climate crisis—and what we can do to avert disaster / Ross
Gelbspan.— 1st ed.
      p.   cm.
   Includes bibliographical references and index.
   HC: ISBN-13 978-0-465-02761-3; 0-465-02761-X
   PBK: ISBN-13 978-0-465-02762-0; 0-465-02762-8
   1. Global warming—Political aspects.   I. Title.

QC981.8.G56G44 2004
363.738'74'0973—dc22

                                                            2004003223

Set in 12-point AGaramond by Perseus Books Group

1 2 3 4 5 6 7 8 9 10—07 06 05

To Tottie—who never bargained for this kind of heaviness when we married thirty years ago and who has responded with so much unqualified support and love and thoughtfulness. Together we marvel at the irony that our personal world overflows with fulfillment and contentment—even as our larger world is darkened by a gathering of intensifying furies.

Contents

Preface to the Paperback Edition

As of this writing in the spring of 2005, New England is suffering through the third coldest May on record—and one of the wettest in the region's history. This cold, wet spring follows one of the most bitter winters in memory. Both, paradoxically, are vivid and immediate consequences of the warming of the planet.

As the Arctic ice melts at a record rate, it releases extremely cold freshwater into the Labrador Current that flows down the coast of northeastern North America. The colder surface waters chill the winds that blow in from the northeast, and they accelerate wind patterns in the far north that drive the Arctic weather fronts southward across the eastern United States.

It is one of the gloomier springs I can recall. The future prospect is not encouraging. New England winters will be different going forward. They will be shorter and far more severe than in the past.

In the year since *Boiling Point* was published, sadly, things have gotten worse. The trends chronicled in the book's first edition progressed without any significant intervention, underscoring the profound difference between the sluggish pace of human, institutional change and the ominous and unpredictable rate of change of the natural world.

To be sure, there is some cause for optimism. In February 2005, the Kyoto Protocol came into effect, albeit without the participation of the United States. The most significant attempt by the U.S. government to begin to regulate carbon dioxide—the McCain-Lieberman Senate bill—may yet survive the 109th congressional session. Several high-profile companies—Cinergy, Exelon, General Electric, among others—have called on the Bush administration to impose emissions limitations and announced they would be lowering their own emissions. The U.S. Department of Energy has announced that the United States could meet goals approximating those of the Kyoto Protocol at a cost of about 15/100ths of one percent of the country's gross domestic product. And 132 U.S. mayors have committed their cities to meet the current low goals of the Kyoto Protocol.

In Europe, following pledges over the past several years of dramatic reductions of coal and oil use by Holland, Germany, and the United Kingdom, French president Jacques Chirac called on the entire industrial world to cut emissions by 75 percent in the next forty-five years.

But all this pales in the face of other developments.

In 2001, when President George W. Bush withdrew the United State from the Kyoto climate negotiations, he declared that the U.S. withdrawal would not affect the efforts of other countries to enact a strong treaty. Nevertheless, in December 2004, when the remaining parties to Kyoto met in Buenos Aires, the United States used its diplomatic leverage to emasculate the next round of climate talks. Although the United States did not ratify the protocol, it was one of the signatories to the 1992 U.N. Framework Convention on Climate Change (UNFCCC) under which the protocol was drafted.

Using its legal leverage under the UNFCC, the United States (together with Saudi Arabia, Kuwait, and other members of OPEC) was able to determine the agenda of the Kyoto discussions. As a result, when the delegates met again in Bonn in May 2005, their deliberations were limited by U.S. intervention to being mere "informational seminars." The Bush administration and its oil-producing allies succeeded in prohibiting delegates from discussing any "action plan" whatsoever. As one climate negotiator observed, despite the fact that the protocol had been ratified by 150 countries, the U.S. action "left the climate talks hanging onto a rock face by their fingernails."

In January, Senator James Inhofe (R-OK) repeated his assertion that climate change "is the greatest hoax ever perpetrated on the American public." His source: *State of Fear*, a book of science fiction based on misinformation provided by a longtime climate "skeptic."

Inhofe's diatribe followed a coal industry initiative to scuttle the efforts of Senators John McCain and Joseph

Lieberman to begin to regulate carbon emissions. An internal coal industry memo, leaked to "Inside EPA," a trade
newsletter published by a private publisher, documented
plans to set up a bogus grass-roots group to bombard senators with letters and e-mails demanding that they not pass
the measure. The same memo contained a plan to publicize
dramatically exaggerated cost estimates of emissions cuts in
order to "sow discord" among a group of states that were part
of a regional compact to cut emissions. And a Canadian mining consultant managed to persuade the *Wall Street Journal* to
run a front-page story on his critique of Michael Mann's critically important reconstruction of the global climate over the
past 1,000 years—a critique that was shown to be baseless by
other climate researchers.

The Bush administration's assault on climate science was
further illuminated in June 2005, when Andrew Revkin of
The New York Times published a front-page story document-
ing how the White House had deliberately misrepresented
the findings of U.S. climate scientists. Revkin reported that
Philip Cooney, a top official of the American petroleum industry before he was tapped by the Bush administration to
head the climate unit of the White House Council on Environmental Quality, had made a number of word changes to
the summary of a scientific document to exaggerate its scientific uncertainties and remove definitive statements about
known impacts of warming.

Within days of that disclosure, Cooney resigned from
his White House post and took a new position with Exxon-
Mobil.

The Bush administration's intransigence toward attempts to address global warming was particularly embarrassing to British prime minister Tony Blair. Although Blair had committed Britain to cutting emissions by 60 percent by 2050, his efforts to secure a strong climate commitment from the Bush administration, his strongest ally in global affairs, prior to a major G-8 meeting in June 2005 fell flat.

Against this background of political gridlock—punctuated by occasional bursts of tepid progress—the pace and intensity of climate change continues to progress.

Four hurricanes, intensified by warming surface waters, devastated parts of Florida in 2004. Altered rainfall patterns resulting in drought have left two million people in Kenya at risk of starvation. A freak December hurricane ravaged Paris and left a quarter of a million homes in France without power. Australian officials imposed stringent water conservation measures in Sydney as a killer drought entered its third year, affecting nearly half the landmass of the country. A record ten typhoons staggered Japan. And 2004 was the fourth hottest year on record worldwide. Such climate impacts cost the global economy about $145 billion in losses during the year.

The scientific community has traditionally been concerned about the gradual impacts of climate change. But in the last year, those concerns have given way to outright worry about *runaway* changes.

The chairman of the U.N.'s Intergovernmental Panel on Climate Change (IPCC), Dr. Rajendra Pachauri, declared in January that we are "passing the point of no return"—and called on the world to make "very deep cuts" in its consumption of carbon fuels "if humanity is to survive." This extremely alarming call by the IPCC chief followed a report in November 2004 by 300 scientists that the Arctic ice cap had shrunk by up to 20 percent over the past 30 years—and that the Arctic, itself, was warming twice as fast as the rest of the world.

Pachauri's remarks reflected the findings of a blue-ribbon international task force on climate change that concluded that, at current rates of emission, the world is ten to twenty years away from an increase of 2 degrees Celsius above the planet's traditional average temperature. Crossing that threshold will likely set in motion a variety of impacts, including a profound change in ocean circulation patterns, the die-off of significant portions of South American rainforests, and a succession of increasingly severe winters that would play havoc with agricultural production in the Northern Hemisphere. The panel included political leaders from the United States, United Kingdom, Australia, China, Germany, and Malaysia, as well as several of the world's leading climate researchers.

Growing numbers of scientists are now looking at the increasing likelihood of a radical and abrupt climate shift. "The weather patterns are changing. The character of the system is changing. It is a signal of how the system is behaving. It is not stable," said Dr. Paul Epstein of Harvard Medical School.

The most common scenario of abrupt climate shift, popularized in the Hollywood blockbuster, *The Day After Tomorrow*, involves a sudden change in ocean current circulation triggered by the infusion of large amounts of fresh meltwater into the dense, salty waters of the North Atlantic. Another scenario involves the rapid release of huge quantities of carbon dioxide from thawing soils in the far north that would trigger a rapid spike in heating.

Although those scenarios, terrifying as they are, are speculative, the documented changes overtaking the planet are not. The earth has become a "net importer" of heat because the buildup of carbon dioxide has altered the traditional balance of the planet which, before the "greenhouse era," radiated as much heat back into space as it absorbed from the sun.

Researchers exploring the depths of the North Atlantic in submarines found that the natural "pump" that pulls the warming current along its traditional path has weakened to about a quarter of its former strength. "Until recently we would find giant 'chimneys' in the sea where columns of cold, dense water were sinking from the surface to the seabed 3,000 meters below, but now they have almost disappeared," said Peter Wadhams, professor of ocean physics at Cambridge University in Britain.

In the year since *Boiling Point* was initially published, nothing has occurred that would alter its urgency. The news media continue to ignore the biggest story in our planet's history. The environmental establishment continues to sugarcoat the truth. A substantial number of religious devotees welcome the accelerating erosion of planetary systems as a certain sign of a coming rapture. And the Bush administration continues to pretend that climate change is of negligible importance.

The solutions that form the centerpiece of the last chapter of the book have survived the scrutiny of critics. The only remaining question is not whether they provide a model for truly effective change but whether or not there is still time to implement them. Absent a sudden, worldwide energy revolution—with all the fundamental changes in our current economic environment that such a transition requires—the answer seems depressingly apparent.

Ross Gelbspan
June 1, 2005

Preface

It is an excruciating experience to watch the planet fall apart piece by piece in the face of persistent and pathological denial.

The situation is reminiscent of *Rhinoceros*, a play written in the 1960s by Eugène Ionesco. The play's main character, Berenger, and his girlfriend, Daisy, watch with dismay as, one after another, their fellow townspeople turn into rhinoceroses. What makes this surrealistic comedy so deeply unsettling is that, except for Berenger and Daisy, the remaining humans in the town refuse to see these transformations. They simply will not acknowledge the fact that, one by one, their colleagues are turning into animals.

This book is a last-gasp attempt to break through the monstrous indifference of Americans to the fact that the planet is caving in around us.

While people in the United States still debate the long-resolved question of whether human-induced climate change is occurring, the overriding challenge for a writer is how honestly and openly to address the real question: Is it already too late to salvage a coherent future?

As a reporter—not an environmentalist—I came to this subject from a very peculiar angle. It began with a collaboration with Dr. Paul Epstein, of Harvard Medical School, on a 1995 article on climate change and the spread of infectious disease in the *Washington Post*.

While researching the piece, I became aware of the enormous dimensions of the climate crisis. This was a subject, I felt, that wanted a much fuller exploration. But after the piece ran in the *Post*, I received several letters from readers who said that, the spread of diseases notwithstanding, they didn't believe the climate was changing. Those readers referred me to the work of a few scientists who were skeptical about climate change.

After reading the writings of some of these "greenhouse skeptics"—Robert Balling's book, *The Heated Debate*, several issues of Pat Michaels's journal, and papers by S. Fred Singer and Richard Lindzen—I was persuaded that global warming was a nonevent. Emotionally, I felt very relieved not to have to deal with such a heavy issue. It was only later that I realized those letters were most likely sent by public relations specialists from the fossil fuel lobby disguised as private readers. At the time, however, I had scheduled interviews with four other scientists, and, as a courtesy, I decided to keep those interviews.

The mainstream scientists showed me how Singer, Michaels, and the other skeptics were manipulating data, selectively omitting critical facts, raising illusory objections, and deliberately misrepresenting the situation. That made me quite angry—and not because I love trees. It made me angry because I had devoted a thirty-one-year career to the belief that in a democracy we need honest information on which to base our decisions. What these few "skeptic" scientists were doing was stealing our reality.

Thinking about the issue, it quickly became clear to me that the climate crisis threatens the very survival of the coal and oil industries—which together constitute one of the biggest commercial enterprises in history. The science is very clear on one point: Climate stabilization requires that humanity cut its consumption of carbon fuels by about 70 percent. That made the motivation behind the disinformation campaign—and the reporting imperative—very clear.

It also led into an amazing drama that continues to unfold just outside the spotlit arena of public awareness.

One goal of my 1997 book, *The Heat Is On*, was to impress on readers the fact that climate change is not a remote, theoretical future risk. Another was to highlight an extraordinarily effective campaign of deception and disinformation by the fossil fuel lobby. Both of those themes recur in this book, but virtually all the developments chronicled here have taken place since 1998.

Sadly, while the particular developments about our changing climate are new, the larger trends are continuing

unabated—and, at least in the United States, essentially unacknowledged.

A large reason for that failure rests with my professional
alma mater—the Fourth Estate. Were the press to make
connections between global climate change and, say, the
thirty-five-inch downpour in Houston two years ago that
left up to $12 billion in damages, the record-setting floods
in northern Europe in the summer of 2002, the recent
increase in tornadoes in the continental interior of the
United States, and the accelerating disintegration of the
planet's ice cover, I think the public would respond. Sadly,
that has not happened.

Another area of the climate issue that has been ignored
by the press involves the ferocious battles on a number of
fronts in which those in the fossil fuel lobby are deploying
huge resources to secure the survival of their industry—even
at the expense of their children's futures.

The most recent iterations of that battle are recounted in
this book. But they began more than a decade ago.

For example, Dr. Benjamin Santer, a world-class climate
modeler at the Lawrence Livermore National Laboratory,
made a huge contribution to our understanding of global
warming with his groundbreaking 1996 study, "A Search for
Human Influences on the Thermal Structure of the Atmosphere." The study profoundly strengthened the case that the
warming of the planet is due to humanity's burning of coal
and oil—not to natural variations in the climate system.
When Santer presented the findings of his research team, he
was subjected to a brutal and life-altering character assault,

highlighted by totally unfounded public accusations of scientific dishonesty by the oil and coal lobby.

Both Santer's experience—and, to a far lesser extent, my own—brought home the bare-knuckle quality of the fight in this country over the reality and the magnitude of the climate crisis.

Shortly after the publication of *The Heat Is On*, the fossil fuel lobby mounted an extensive campaign accusing me of résumé fraud. They circulated a message on the Internet and elsewhere that I had falsely claimed to be a co-recipient of a Pulitzer Prize. My reaction was mixed. On the one hand, it was quite hurtful. On the other hand, I was privately pleased. The fact that the lobbyists for big coal and big oil had to resort to a campaign of character assassination meant there was nothing in the book with which they could find serious fault.

For the record, the Pulitzer grew out of a series of articles I conceived and edited as special projects editor at the *Boston Globe* about twenty years ago. The project revealed a systematic pattern of job discrimination against African Americans in every major sector of Boston's economy. A Pulitzer Prize technically goes to the named reporters and the newspapers. In this case, because I had conceived the series, helped select the reporters, directed the reporting, and edited the articles, the editor and publisher of the *Globe* designated me a co-recipient of the prize on behalf of the newspaper. The *Globe* published my photo and bio, along with the other members of the team, under the headline "Pulitzer Prize Winners." The newspaper's board of directors and the then mayor of

Boston congratulated me for my role in the project. Those documents, along with several others posted on my Web site, http://www.heatisonline.org, verify my role in that prize. Personally, I was quite proud of the contribution the project made to the civic life of Boston.

The hope underlying this book is of a different type. Rather than striving to correct injustices, this book is intended to illuminate an extraordinarily positive potential that is obscured by the deeply fearful prospect of runaway climate change. Even while growing numbers of people around the world now acknowledge the reality of global warming, the vast majority continue to minimize its urgency and scale. I hope the glimpses of what is currently happening to the planet, spotted throughout the book, will convey a sense of the speed with which we must address this crisis.

But the driving motivation behind this book is to change the way we regard this threat—to make our responses far less fearful and defensive and to recast this moment in human history as an unprecedented opportunity to use the climate crisis as the foundation for a historically unprecedented common global project that would create the basis for a far wealthier, more equitable, and ultimately, much more peaceful world. A real solution to the climate crisis could, if properly structured, provide a liftoff toward a far more promising human future.

Ross Gelbspan
January 14, 2004

Acknowledgments

I owe great thanks to far too many people to acknowledge here. The enthusiasm of climate activists around the country of all ages—many of them responding to the moral and ethical challenges of the climate crisis—has been a buoyant source of inspiration. It is a frequent temptation, in working on this story, to succumb to depression or terminal frustration. The antidote for me has come from the examples of so many political and religious activists all over the country who are giving so much of their energy and determination to this battle.

On a more personal level, I have drawn huge amounts of support and encouragement from several friends. Dianne Dumanoski, a longtime colleague and world-class writer and reporter, has deflected my forays into emotional self-indulgence with her sometimes frightening intellectual honesty. Whatever climate catastrophes occur, she has reminded me, there will still be people on this planet—no matter how fragmented and impoverished and degraded their existence—who may benefit from clues about how to begin again. Paul Epstein has consistently shared with me his scientific insights and his personal intuitions. His understanding of the dynamics and interactions of systems has more than helped shape my thinking. It has been an indelible intellectual gift. But perhaps of even more value has been his relentless insistence on the positive potential that a real solution contains for humanity. With his periodic reminders that large-scale social change can erupt as suddenly as rapid climate change, he has refused to allow the most discouraging signals to cripple his spirit—or mine. Finally, my friend Dick Russell, a writer of extraordinary talent and prolific output, continues to point out that writing, itself, is an act of faith and the mere act of doing it assumes a future.

In addition to these very special friends, I am particularly grateful to my agent and my editor. Ike Williams, my agent, has been unflaggingly supportive and enthusiastic, even finding a magical way to expedite the publication of this book. Amanda Cook, my editor at Basic Books, has been invaluable in snapping a disjointed and muddled manuscript into what I hope is a clear and coherent book. Amanda must be the type of person James Joyce had in mind when he wrote that he was searching for "the ideal reader with the ideal insomnia."

Finally, my most heartfelt gratitude goes to my daughters, Thea and Joby. For the past few years, they have been hounding me to give up this work and take up some other project that is less draining and probably more age-appropriate. I sense they are trying to protect me from the disappointment of the likely failure of my own quixotic dream. But they may not completely share the terrible knowledge of what this world will be like in another thirty-five years when they are my age. And besides, neither of them is a father.

Boiling Point

1

Not Just Another Issue

We've known for some time that we have to worry about the
impacts of climate change on our children's and grandchildren's
generations. But we now have to worry about ourselves as well.
✍ MARGARET BECKETT,
BRITISH SECRETARY OF STATE FOR ENVIRONMENT
APRIL 26, 2002

By late 2003, the signals were undeniable: Global climate
change is threatening to spiral out of control.

The six-month period from June to December 2003
brought a succession of scientific findings, climate impacts,
political and diplomatic developments, and responses from
the financial world that vividly underscored the urgency and
magnitude of the climate crisis.

The events of that year surprised even many seasoned
climate scientists—and brought home to many others the fact
that, given all its ramifications, the climate crisis is far more
than just an environmental issue. It is a civilizational issue.

Nevertheless, by the end of 2003, most Americans were still in denial.

The evidence is not subtle. It is apparent in the trickling meltwater from the glaciers in the Andes Mountains that will soon leave many people on Bolivia's mountainside villages with no water to irrigate their crops and, after that, not even enough to drink. It is visible in the rising waters of the Pacific Ocean that recently prompted the prime minister of New Zealand to offer a haven to the residents of the island nation of Tuvalu as it slowly goes under. It is evident in the floods that, in 2002, inundated whole cities in Germany, Russia, and the Czech Republic. It is underscored in the United States by the spread of West Nile virus to forty-two states—and to 230 species of birds, insects, and animals—and in the record-setting 412 tornadoes that leveled whole towns during a ten-day span in May 2003. Its reality is visible from outer space—where satellites have detected an increase in the radiation from greenhouse gases—to our own backyards.

Seen in its full dimensions, the challenge of global climate change seems truly overwhelming. In the absence of a compelling and obvious solution, the most natural human tendency is simply not to want to know about it.

When a crisis becomes so apparent that denial is no longer tenable, the typical response is to minimize the scope of the problem and embrace partial, inadequate solutions. Witness the voluntary approach of the Bush administration as well as the low goals of the Kyoto Protocol, which calls for industrial countries to cut their aggregate emissions by

5.2 percent below 1990 levels, by 2012. (The goal for the United States under the treaty was reductions of 7 percent below 1990 levels.)

By contrast, the science is unambiguous: To pacify our increasingly unstable climate requires humanity to cut its use of coal and oil by 70 percent in a very short time. The grudging response in the United States, and to a lesser extent, abroad, reflects more than a profound underestimation of the scope and urgency of the problem. It betrays an equivalent underestimation of the truly transformative potential of an appropriate solution. Given the scope of the challenge, a real solution to the climate crisis seems to offer a historically unique opportunity to begin to mend a profoundly fractured world.

But it all begins with the climate—and the stunningly rapid atmospheric buildup of carbon dioxide emissions from our fossil fuels. This is trapping growing amounts of heat inside our atmosphere, heat that has historically radiated back into space.

Unintentionally, we have set in motion massive systems of the planet (with huge amounts of inertia) that have kept it relatively hospitable to civilization for the last 10,000 years. With our burning of coal and oil, we have heated the deep oceans. We have reversed the carbon cycle by more than 400,000 years. We have loosed a wave of violent and chaotic weather. We have altered the timing of the seasons. We are living on an increasingly precarious margin of stability.

The accelerating rate of climate change is spelled out in two recent studies—one on the environmental side, one on the energy side.

In 2001, researchers at the Hadley Center, Britain's main climate research institute, found that the climate will change 50 percent more quickly than was previously assumed. That is because earlier computer models calculated the impacts of a warming atmosphere on a relatively static biosphere. But when they factored in the warming that has already taken place, they found that the rate of change is compounding. Their projections show that many of the world's forests will begin to turn from sinks (vegetation that absorbs carbon dioxide) to sources (vegetation that releases carbon dioxide)— dying off and emitting carbon—by around 2040.

The other study, from the energy side, is equally troubling. Three years ago, a team of researchers reported in the journal *Nature* that unless the world is getting half its energy from noncarbon sources by 2018, we will see an inevitable doubling—and possible tripling—of atmospheric carbon levels later in this century. In 2002, a follow-up study by many of the same researchers, published in the journal *Science*, called for a Manhattan-type crash project to develop renewable energy sources—wind, solar, and hydrogen fuel. Using conservative estimates of future energy use, the researchers found that within fifty years, humanity must generate at least three times more energy from noncarbon sources than the world currently produces from fossil fuels to avoid a catastrophic buildup of atmospheric CO_2 later in this century.

For nearly a decade after it surfaced as a public issue in 1988, climate change was regarded primarily as a remote, almost futuristic, threat based on an arcane branch of science that depended on the mind-numbing complexity and paralyzing uncertainty of an early generation of computer models whose reliability was too suspect to justify enormous policy changes.

In 1995, the issue gained prominence when the world's community of climate scientists first declared they had detected the "human influence" on the climate. That finding legitimized global climate change as a major environmental issue. As a consequence, climate change was subsequently accorded the same mix of rhetorical concern and political inaction as most other environmental issues.

In 2001, however, the issue was infused with a jolt of urgency. That January, the U.N. Intergovernmental Panel on Climate Change (IPCC) concluded that the climate is changing far more rapidly than scientists had previously projected.

More than 2,000 scientists from 100 countries, participating in the largest and most rigorously peer-reviewed scientific collaboration in history, reported to the UN that brutal droughts, floods, and violent storms across the planet will intensify because emissions from humanity's burning of coal and oil is driving up temperatures much more rapidly than scientists had anticipated just six years earlier.

"The most comprehensive study on the subject [indicates] that Earth's average temperature could rise by as much as 10.4 degrees over the next 100 years—the most rapid change in 10 millennia and more than 60 percent higher than the same group predicted less than six years ago," according to the *Washington Post*.

Rising temperatures will melt ice sheets and raise sea levels by as much as thirty-four inches, causing floods that could displace tens of millions of people in low-lying areas— such as China's Pearl River Delta, much of Bangladesh, and the most densely populated area of Egypt.

Droughts will parch farmlands and aggravate world hunger. Storms triggered by such climatic extremes as El Niño will become more frequent. Diseases such as malaria and dengue fever will spread, the report noted.

A second working group of the IPCC—one that focused on the impacts of coming climate changes—reached the extremely sobering conclusion that "most of earth's inhabitants will be losers," in the words of the group's co-chair, James McCarthy of Harvard University.

The report concluded that poor countries in Africa, Asia, and Latin America with limited resources would bear the brunt of the most extreme climate changes. It added that economic losses from natural catastrophes increased from about $4 billion a year in the 1950s to $40 billion in 1999, with about one-fourth of the losses occurring in developing countries.

(Two years later, nature had already upped the ante. In 2003, the United Nations reported that climate impacts cost the world $60 billion that year, an increase of 10 percent over the $55 billion in climate-related damages in 2002.)

"The scientific consensus presented in this comprehensive report about human-induced climate change should sound alarm bells in every national capital and in every local community. We should start preparing ourselves," declared

Klaus Topfer, director of the United Nations Environment Programme (UNEP).

In the fall of 2003, a succession of events—climatic, economic, and political—coalesced into a vivid mosaic that reflects the reach and variety of climate impacts and their reverberation through our economic and political institutions.

Several developments, which are examined in more detail later in this book, were particularly ominous because of their scope:

- The entire ecosystem of the North Sea was found to be in a state of collapse because of rising water temperatures.
- For the first time in recorded history, the world consumed more grain than it produced for *four years in a row*. The reason: rising temperatures and falling water tables—both consequences of global climate change.
- The German government declared that the goals of the Kyoto Protocol need to be increased by a factor of four to avoid "catastrophic" changes. Otherwise, the climate will change at a rate not seen in the last million years.
- The most highly publicized impact of global warming in 2003 involved a succession of headlines from Europe about an extraordinary summertime heat wave. Scientists attributed the unusually high mortality rates not to the fact that the August temperatures were so much higher than before. The

record-setting temperatures provided only a partial explanation. The link between climate change and the deaths of so many Europeans had been established in a laboratory of the U.S. National Oceanic and Atmospheric Administration (NOAA) nearly six years earlier, when researchers at NOAA's National Climatic Data Center found that, as Earth's temperature has been rising, the nighttime low temperatures have been rising nearly twice as fast as the daytime high temperatures. Before the buildup of heat-trapping carbon dioxide in the atmosphere, daytime and nighttime temperatures generally rose and fell in parallel. But as carbon levels in the atmosphere have thickened, they have tended to trap heat during the evenings, preventing it from radiating back into space once the sun has faded into the nighttime sky.

In August 2003, that finding took on an especially grisly reality. The lingering nighttime warmth in Europe that summer deprived overheated Europeans of the normal relief from blistering daytime temperatures. As a result, people were not able to recuperate from the heat stress they had suffered during the relentlessly hot days. When that brutal summer finally subsided, it left more than 35,000 people dead.

- The following month, silently and out of view of most of the world, the biggest ice sheet in the Arctic—3,000 years old, 80 feet thick, and 150 square miles in area—collapsed from warming surface waters in September 2003. The Ward Hunt Ice Shelf, located 500 miles from the North Pole on the edge of Canada's Ellesmere Island, broke in two. A massive freshwater lake long held back by the ice also drained away.

- The same month brought another startling—and largely unanticipated—consequence of our fossil fuel use. Scientists reported that the pH level of the world's oceans had changed more in the last 100 years than it had in the previous 10,000 years—primarily because of the fallout from emissions caused by coal and oil burning. In short, the oceans are becoming acidified.

- By the fall of 2003, an eighteen-month drought in Australia had cut farm incomes in half—and left many scientists speculating that the prolonged drought may have become a permanent condition in one of the country's richest food-growing areas.

Nor was it only the planet's physical systems that felt the threat of escalating climate change. Many financial institutions also began to feel the heat in late 2003.

Pension fund managers, bankers, and Wall Street advisers—representing more than $1 trillion in assets—issued a "call to action" in November 2003 about impending climate-driven upheavals in the world's financial markets. At the meeting, which was sponsored by the United Nations, the treasurer of the state of California, Philip Angelides, declared: "In global warming, we are facing an enormous risk to the U.S. economy and to retirement funds—that Wall Street has so far chosen to ignore."

The meeting was the most elaborate effort yet by a growing group of fund managers and other finance officers to "persuade businesses to move more aggressively to identify and address problems they might face from global warming, increasingly frequent extreme weather and other climate changes that have been linked to the rapid buildup in the atmosphere of carbon dioxide and other heat-trapping gases," according to the *New York Times*.

Perhaps the strongest response of the financial community to the climate crisis came from one of the world's largest insurers. In May 2003, Swiss Reinsurance announced that it was asking directors and officers of its client companies what their firms were doing to reduce their use of fossil fuels. The company made clear that if those corporate officials were not moving aggressively enough to reduce their carbon emissions, they would risk losing liability insurance.

But if the increasingly visible risks were causing ripples in the financial world, they seemed the subject of an almost perverse kind of satisfaction by the world's largest oil company, ExxonMobil.

In late 2003, the oil giant announced it was anticipating a 50 percent increase in global carbon emissions by the year 2020. "Between now and 2020 we estimate increases of some 3.5 billion tonnes per year of additional carbon emissions, so it's definitely increasing," said Randy Broiles, global planning manager for ExxonMobil's oil and gas production unit. Despite expected increases in energy efficiency, more cars, rising industrial output, and rising living standards in the developing world will create a worldwide demand for about 40 percent more energy in the next two decades, Broiles said.

Unlike the world's community of climate scientists and most of the world's governments, ExxonMobil stands to benefit from the coming surge in carbon emissions. "The oil resource base is huge—it's huge," Broiles told a conference, "and we expect it to satisfy world demand growth well beyond 2020."

The final—and perhaps most disheartening—piece of climate news from the fall of 2003 came from the Kremlin when a top deputy to President Vladimir Putin declared that Russia was withdrawing from the Kyoto Protocol. As delegates began yet another round of climate negotiations in Milan at the beginning of December 2003, one of Putin's top economic advisers announced that Russia would withdraw from the Kyoto process because it was not in the country's economic interest.

With the United States already having withdrawn from the negotiations, Russia's defection—if it proved permanent—would drive the final stake through the heart of the climate treaty, which requires ratification by fifty-five countries that collectively generate 55 percent of the world's carbon emissions in order to take effect. (The United States accounts for about 25 percent of those emissions.) Some observers believed the Russians were simply using the threat to secure more favorable treatment under the Kyoto process. Others pointed out that Russia had already secured very generous emissions allowances under the treaty. They speculated that Russia threatened to withdraw from Kyoto because the prior withdrawal of the United States had deprived Russia of a major buyer for the emissions credits it had been granted under the treaty. In any event, the Russian vacillations demoralized delegates from other countries that were party to the near moribund Kyoto process, which was already teetering on the brink of irrelevance.

Ironically, the year 2003 also seemed to mark a sea change among many segments of American society on the climate issue. The failure of the world's diplomats to implement meaningful solutions has generated a groundswell of grassroots and voluntary climate action around the country. Some thirty states—and more than 100 U.S. cities—have initiated plans to reduce their own carbon emissions. Hundreds of colleges and universities have embarked on "green campus" programs. Growing numbers of religious leaders and their congregations are responding to what they see as the ethical challenges embedded in climate change. A number of large environmental groups have taken up the climate as their central issue. And grassroots activist groups dedicated to the climate crisis have sprung up all over the country.

But the enthusiasm of many of these groups far exceeds both their resources and their political influence. Many activists today are promoting the use of energy-efficient lightbulbs, carpooling, and other climate-conscious lifestyle changes. Unfortunately, these strategies, even under the most wildly optimistic scenarios, fall far short of nature's requirement that we cut our consumption of coal and oil by 70 percent.

More to the point, they are desperately outmatched by the financial power and political influence of big coal and big oil in what basically boils down to a titanic conflict of interests over the future of this civilization.

What began as a normal business response by the fossil fuel lobby—denial and delay—has now attained the status of a crime against humanity. The coal and oil industries in the United States have scuttled every substantive effort to change our lethal energy diet in the interest of preserving their companies' short-term gains.

By the end of 2003, it was clear that the fossil fuel lobby enjoyed not only immense financial resources. It also had the unwavering support of arguably the most powerful political office in the world—the White House of President George W. Bush.

The coming chapters of this book will focus less on the science of global warming and more on its dimensions in our social and political lives.

Chapter 2, "The Sum of All Clues," is a capsule version of the most important responses of the world's community of climate scientists to the central question underlying the climate crisis: How do we know this upheaval is caused by human beings rather than resulting from the wild natural swings that have marked the prehistoric record of the global climate?

Chapter 3, "Criminals Against Humanity," documents an extraordinary collaboration between the fossil fuel industry and the White House to keep this issue out of public view in the United States—and to keep the debate focused on whether it is happening rather than on what to do about it.

Chapter 4, "Bad Press," examines, from another angle, the reason the American public is so ill-informed about climate change compared to the rest of the world. Part of the answer lies in some outdated journalistic conventions that were adopted to ensure objectivity but which, in fact, have generated frequently inaccurate and, in some cases, grossly distorted portraits of the state of our scientific knowledge of what is happening to the planet.

Chapter 5, "Three Fronts of the Climate War," details divisions on the climate crisis: between the United States and much of the rest of the world, between Washington and much of the rest of the country, and within the oil, auto, and insurance industries.

Chapter 6, "Compromised Activists," examines the first, groping attempts of the world to deal with climate change—and the extraordinary frustration of thousands of political, religious, and campus-based climate activists who find themselves forced by a steel wall of denial to lower their expectations and pursue only the most minimal goals.

Chapter 7, "Thinking Big: Three Beginnings," portrays (I hope as honestly and sympathetically as possible) three large-scale proposals that, while flawed in this author's eye, stand out for their intellectual courage in trying to address the true scope and scale of the challenge.

Chapter 8, "Rx for a Planetary Fever," details my own preferred proposal—a set of three interactive global-scale policy strategies, which are designed to address the climate crisis but which could also generate many other positive changes in our social, political, and economic lives.

Interspersed between the chapters are snapshots of warming-driven impacts on the systems of the planet. These impacts are not the products of computer-model projections, nor are they examples of increasingly violent, climate-driven weather extremes. They are documented physical changes that are taking place as you read this.

But the real news about climate change is not about its destructive potential. The real news lies in the extraordinary opportunity the climate crisis presents to us. Given how very central energy is to our existence, a solution to climate change—which is appropriate to the magnitude of the problem—could also begin to reverse some very discouraging and destructive political and economic dynamics as well.

Nature's requirement that humanity cut its use of carbon fuels by 70 percent in a very short time leaves us with basically two choices. We can either regress into a far more primitive and energy-poor lifestyle—or we can mount a global project to replace every oil-burning furnace, every coal-fired generating plant, and every gasoline-burning car with noncarbon and renewable energy sources. A properly framed plan to rewire the globe with solar energy, hydrogen fuel cells, wind farms, and other sources of clean energy would do much more than stave off the most disruptive manifestations of climate change. Depending on how it is structured, a global transition to renewable energy could create huge numbers of new jobs, especially in developing countries. It could turn dependent and impoverished countries into robust trading partners. It could significantly expand the overall wealth in the global economy. It could provide many of the earth's most deprived inhabitants with a sense of personal future and individual purpose.

That same solution also contains the seeds for real security in a world that threatens to become polarized between totalitarians and terrorists.

It could very well trigger a major change in many unsustainable practices that are threatening many other natural systems—the world's forests and oceans, for example—whose vitality is already at risk.

It could be the prompt that reverses the kind of exaggerated nationalism that threatens to re-tribalize humanity. Rather than hastening our regression into a more splintered, combative, and degraded world, it could be the springboard that propels us forward into a much more cooperative and coordinated global community.

The same solution to the climate crisis could also begin to put people in charge of governments and governments in charge of corporations. A program to rewire the world with clean energy could provide the outlines of a model in which people would put democratically determined boundaries around the operations of multinational corporations.

In the long run, the solutions to the climate crisis could establish equity as a universal human value and resuscitate participatory democracy as a governing operating principle that reorganizes our relationships to each other, to other nations, to the global economy, and most fundamentally, to the planet on which we all depend.

In this immense drama of uncertain outcome, this much is true: A major discontinuity is inevitable. The collective life we have lived as a species for thousands of years will not continue long into the future. We will either see the fabric of civilization unravel under the onslaught of an increasingly unstable climate—or else we will use the construction of a new global energy infrastructure to begin to forge a new set of global relationships.

If we are truly lucky—and visionary enough—those new relationships will differ dramatically from what we have known throughout our recorded history. They will be based far less on what divides us as a species and far more on what unites us. Embedded in the gathering fury of nature is a hidden gift—an opportunity to begin to redeem an increasingly fragmented world.

The alternative is a certain and rapid descent into climate hell.

SNAPSHOTS OF THE WARMING NO. I

In a time of war, truth is the first casualty.

In a time of warming, it begins with the ice.

That is followed in short order by plants, animals, fish, birds, and entire ecosystems that are right now migrating toward the poles all over the world in the futile pursuit of stable temperatures.

The next set of casualties is people.

Only the bugs love warming.

*A*lthough the most obvious signal of early-stage global warming involves the increase in extreme weather, the most strikingly visible impact of climate change lies in the melting away of Earth's ice cover.

Scientists had documented the thinning of Arctic sea ice, but they were shocked by the sudden rupture of the largest ice shelf in the Arctic, the Ward Hunt Ice Shelf, in September 2003. "Large blocks of ice are moving out. It's really a breakup," Warwick

Vincent, a professor of biology at Laval University in Quebec, told the Los Angeles Times. He added: "We'd been measuring incremental changes each year. Suddenly in one year, everything changed."

Three larger ice shelves had broken off Antarctica since 1995, but this was the largest ice separation in the Arctic. It occurred, moreover, in an area of the eastern Arctic long thought to be relatively insulated against the gradual warming of the planet.

"This type of catastrophic [event] is quite unprecedented," said Martin Jeffries, a professor of geophysics at the University of Alaska–Fairbanks. Jeffries noted that the Arctic ice has thinned dramatically over the last twenty years. In 1980, the same ice shelf was 150 feet thick. By 2003, it had thinned to less than half that depth.

That breakup follows the disintegration of three large pieces of Antarctic ice shelf since 1995. The latest piece, about the size of Rhode Island, collapsed in just thirty-one days in the spring of 2002. For years, researchers had been watching pieces of the Larsen Ice Shelf B slowly break away, but the speed of disintegration was "staggering," according to David Vaughan, a glaciologist at the British Antarctic Survey. Added Pedro Skvarca, director of glaciology at the Antarctic Institute of Argentina: "We are witnessing a very significant warning sign of climate warming."

Because ice shelves float on the surface of the ocean, they do not displace water when they rupture. But a major, and potentially very rapid, rise in sea levels could result from the collapse of huge land-based glaciers like the Greenland Ice Sheet. That glacier—the largest on the planet—has, since 1993, been losing *three cubic miles of ice a year, which is enough to cover the state of Maryland with a foot of ice. Recently, scientists expressed concerns about the stability of the ice sheet—after discovering that meltwater from its surface was trickling down to its base, lubricating the bedrock on which it sits. Were the Greenland Ice Sheet (or a substantial portion of the West Antarctic Ice Sheet) to slide into the oceans, it could cause a rapid rise in sea levels. Since about half the world's population lives near coastlines, the consequences could be chaotic.*

But even absent this catastrophic scenario, the rapid melting of earth's ice cover is startling.

A National Aeronautics and Space Administration (NASA) study in late 2002 found that the ice cover on the Arctic Ocean is vanishing at an astonishing rate. About 9 percent of that ice cover is vanishing every decade—about three times faster than scientists had predicted.

For example, the Arctic ice shelf that ruptured in the fall of 2003 had been found to be 150 feet thick in 1980. When it broke in two, spawning icebergs throughout the area, it had melted to about half that thickness.

In the Andes, the glaciers are melting far more rapidly than anyone had expected. Glaciers around Chacaltaya, the highest mountain in Bolivia, for instance, are retreating by about thirty feet a year. Rural mountain communities in Peru, Ecuador, Bolivia, and Colombia are already losing water for both irrigation and drinking.

Glacier National Park in Montana featured about 150 glaciers in the nineteenth century, when naturalists hiked the park. A century later, there are thirty-five. The cold slivers that remain are disintegrating so fast that scientists estimate the park will have no glaciers in thirty years.

In Kenya, scientists project that the famed ice cap on Mount Kilimanjaro will be completely gone in twenty years.

In India, scientists estimate that the glaciers in the Himalayas, the highest mountain range on earth, could disappear in another thirty years.

2

The Sum of All Clues

Prehistoric and early historic societies—from villages to states to empires—were highly vulnerable to climatic disturbances. Many lines of evidence now point to climate as the primary agent in repeated social collapse.

✍ ANTHROPOLOGIST HARVEY WEISS AND PALEOCLIMATOLOGIST RAYMOND S. BRADLEY, "WHAT DRIVES SOCIETAL COLLAPSE," *Science*
JANUARY 26, 2001

With the impacts of climate change becoming so visible and so ominous—melting ice, rising sea levels, increasingly violent weather, and changes in the timing of the seasons— the fossil fuel industry, and the so-called greenhouse skeptics they fund, can no longer deny the reality of the heating of the planet.

Over the past ten years, the arguments against the reality of climate change by the carbon lobby have been as inconsistent as the weather itself. During the early years of the 1990s, the fossil fuel lobby insisted that global warming was not happening. In the face of incontrovertible findings by the scientific community, the fossil fuel industry then conceded climate change is, indeed, happening but that it is so inconsequential as to be negligible. When new findings indicated the warming is, indeed, significant, the spokespeople for the coal and oil industries then put forth the argument that global warming is good for us.

But the throughline of the industry's various arguments has centered on the fact that the global climate has been marked throughout Earth's history by large swings and periods of great instability. The central argument that big coal and big oil have spent millions of dollars to amplify over the last decade is that the warming is a natural phenomenon on which human beings have little or no impact.

That argument was first answered in 1995 by the world's community of climate scientists when they determined that the warming is, undeniably, due to human activities. Since that 1995 declaration, a succession of new findings has strengthened the case for human-induced climate change beyond a doubt. This is not the hypothesis of a few researchers. The finding that human beings are changing the climate comes from more than 2,000 scientists from 100 countries reporting to the United Nations in what is the largest and most rigorously peer-reviewed scientific collaboration in history.

In 1988, the United Nations created the Intergovernmental Panel on Climate Change to find out why the planet is warming: Was it attributable to the natural variability of the climate, or was it due to human activities?

Seven years later, the IPCC declared that the scientific panel had found—in the conservative language of science—a "discernible human influence" on the climate. After reviewing the scientific literature on climate change, the IPCC found that the heating of the planet was due to the buildup of greenhouse gases—primarily carbon dioxide from our burning of coal and oil—in our atmosphere.

That 1995 consensus declaration was based on a number of findings, including three critical research efforts.

That year, a team of researchers led by Dr. Benjamin Santer of the Lawrence Livermore Lab examined the pattern of heating in the atmosphere. That pattern of warming—over land and water and warm and cold areas—produced a very specific pattern, one that matches the pattern projected by computer models of "greenhouse gas," plus sulfate warming. When the vertical structure of the warming was examined, it was found to be graphically different from the structure produced by natural warming.

A second "smoking gun" was published in 1995 when a team of scientists at NOAA's National Climatic Data Center verified an increase of extreme weather events in the United States. They concluded the growing weather extremes are due, by a probability of 90 percent, to rising levels of greenhouse gases. Those extremes—which reflect an intensification of the planet's hydrological cycle from atmospheric heating—are not consistent with natural warming and, instead, resemble the changes that were projected for emissions from fossil fuels. The researchers declared the climate in the United States is becoming more "greenhouse-like"—with more intense rain and snowfalls, more winter precipitation, more droughts, floods, and heat waves. It concluded: "[T]he late-century changes recorded in U.S. climate are consistent with the general trends anticipated from a greenhouse-enhanced atmosphere."

A third contribution to our understanding of the global climate appeared that same spring when David J. Thomson, a signals analyst at AT&T Bell Labs, evaluated a century of summer and winter temperature data. Whereas some scientific skeptics had attributed this century's atmospheric warming to solar variations, Thomson discovered the opposite: The accumulation of greenhouse gases had overwhelmed the relatively weak effects of solar cycles on the climate. He also discovered that since the beginning of World War II, when accelerating industrialization led to a skyrocketing of carbon dioxide emissions, the timing of the seasons had begun to shift. Since 1940, he wrote in the journal *Science*, the seasonal patterns "of the previous 300 years began to change and now appear to be changing at an unprecedented rate."

Since the IPCC's 1995 declaration, a succession of studies has profoundly strengthened the case for human-induced global warming.

In 1997, a research team led by David Easterling of NOAA's National Climatic Data Center found that the nighttime and wintertime low temperatures are rising nearly twice as fast as the daytime and summertime high temperatures. Easterling called the findings a "fingerprint" study of "greenhouse warming." That research, based on data from

5,400 observing stations around the world, showed that "[t]he rise in [minimum temperatures] is due to higher humidity and more water vapor, especially in the winter in northern latitudes of the Northern Hemisphere. In an increasingly 'greenhouse' world this is the kind of rise you'd expect to see," Easterling wrote. He added that if the warming were natural, and not driven by fossil fuel emissions, the high and low temperatures would more or less rise and fall in parallel.

In 1999, a team of British meteorologists studied all the factors that influence changes in the climate—solar activity, volcanic eruptions, emissions of sulfur particulates, and greenhouse gases. According to an editorial in the journal *Nature,* "The researchers' findings were unambiguous: 'The temperature changes over the Twentieth Century cannot be explained by any combination of natural internal variability and the response to natural forcings alone,' they conclude. Rather, it seems necessary to include some human-induced component in the climate forcing throughout the century . . . ."

"Thus the rise in temperature of about a quarter of a degree since the 1940s seems to be due mainly to increases in greenhouse gases . . . All in all," the editorial concluded, "it seems we can lay to rest the idea that recent climate warming is just a freak of nature."

A year later, Thomas Crowley at Texas A&M University concluded that 75 percent of the warming of the twentieth century was due to human activities—and that the rate of warming exceeded any similar time period in the last 1,000 years.

The Crowley findings affirmed a groundbreaking study published two years earlier by a team of scientists who essentially reconstructed the history of the global climate over the previous 1,000 years. Michael Mann, Raymond Bradley, and Malcolm Hughes published their findings in the peer-reviewed journal *Geophysical Research Letters*. The reconstruction involved examinations of tree rings, ice cores, and sediments that contain information about earlier climatic periods.

They found that since the year 1000, the decade of the 1990s was the hottest in history—and that 1998 was the warmest year at least in the millennium. Their research, captured in a famous "hockey-stick" graph, showed that from about the year 1,000 to the mid-nineteenth century, the climate was actually cooling very slightly—about one-fourth of a degree. But in the last 150 years, beginning with the widespread industrialization of the late nineteenth century, the temperature has shot upward at a rate unseen in the last 10,000 years.

Those studies—which used computer models and physical climate "archives"—were corroborated for the first time by direct evidence from satellites in 2001.

That year, scientists studying the escape of longwave radiation from Earth into the outer atmosphere discovered there had been a marked change between 1970 and 1997. Using data gathered by satellites in those two years, the scientists found that radiation from greenhouse gases had increased significantly over the twenty-seven-year period.

The satellite radiation readings, according to researchers, provided the first direct experimental evidence "for a significant increase in the Earth's greenhouse effect that is consistent with concerns over radiative forcing of climate."

Those concerns were heightened in 2000 when scientists determined that the rate of heating had skyrocketed in the latter part of the twentieth century.

In early 2000, scientists declared that the earth's surface is warming at an "unprecedented rate" that was not expected to be seen until well into the twenty-first century.

According to an analysis by a team of climatologists, led by Tom Karl of the National Oceanic and Atmospheric Administration, although warming for most of the twentieth century was progressing at the rate of 1°C per century, that changed in the mid-1970s. Since 1976, the earth has been warming at a rate of about 3°C per century. Karl speculated that the planet may have experienced a "change point" at which the rate of warming suddenly accelerated.

Said Jonathan Overpeck, director of the University of Arizona's Institute for the Study of Planet Earth: "There is no known precedent of natural forces that could have given rise to the temperatures of the last decade."

That heating was apparent not only in atmospheric studies but in measurements of the deep oceans as well. In 2001, two studies indicated that the warming was penetrating to far deeper levels—with potentially irreversible consequences.

That year, two teams of researchers, one headed by Sydney Levitus of the National Oceanic and Atmospheric Administration, and the other by Tim Barnett of the Scripps Institute of Oceanography, found that the world's oceans had warmed by about one-tenth of 1°C down to a depth of

3,000 meters—almost two miles—in the last fifty years. Said Levitus: "I believe our results represent the strongest evidence to date that the Earth's climate system is responding to human-induced forcing."

Levitus and his colleagues calculated the average of how much the oceans had warmed by compiling millions of deep ocean temperature measurements from 1948 through 1995. But initially they couldn't say for sure whether the heat came from greenhouse warming or from a natural swing in the climate cycle.

To solve that riddle, Levitus and Barnett each used a different computer model of the earth's climate to simulate how ocean temperature should respond to current levels of greenhouse gases and other modern atmospheric conditions. The amount of warming predicted by both models matched the warming that had been physically measured.

The Scripps team ran their model with—and without—the extra greenhouse gases and sulfate aerosols produced by combustion of coal and oil. "What we found is that the signal is so bold and big that you don't have to do any fancy statistics to beat it out of the data. It's just there, bang," said Dr. Barnett. He added that the findings "will make it much harder for naysayers to dismiss predictions from climate models."

The findings of the Levitus team also answered a major question posed by "greenhouse skeptics." The skeptics contended that if the climate models were accurate, the atmosphere should have warmed considerably more than it has. But the findings from deep ocean measurements showed that a substantial portion of that heat had been absorbed by the world's oceans. "We've shown that a large part of the 'missing warming' has occurred in the ocean," said Levitus, who added: "The whole-Earth system has gone into a relatively warm state."

By 2003, the science had become so robust that even the most conservatively spoken scientists were adamant about the fact that humans, primarily through their burning of fossil fuels, are heating the atmosphere.

"Modern climate change is dominated by human influences, which are now large enough to exceed the bounds of natural variability. The main source of global climate change is human-induced changes in atmospheric composition . . . Anthropogenic climate change is now likely to continue for many centuries. We are venturing into the unknown with climate, and its associated impacts could be quite disruptive," wrote Thomas Karl and Kevin Trenberth in the journal *Science*.

Putting the various strands of temperature research together, the picture that emerges is profoundly ominous.

It begins with bare physical measurements—independent of computer models. Carbon dioxide in the atmosphere traps in heat. For the last 10,000 years, the amount of CO_2 remained constant at about 280 parts per million—until the late nineteenth century, when the world began to industrialize using coal and oil. Today, that 280 is up to 379 parts per million. That is a level the planet has not experienced for 420,000 years.

The most direct consequence of this buildup of atmospheric carbon is in the relentless rise of global average temperatures.

Seventeen of the eighteen hottest years on record have occurred since 1980. The period from 1991 to 1995 constitutes the hottest five-year period on record. The year 1998 replaced 1997 as the hottest year in human history, and 2001 replaced 1997 as the second-hottest year. Then 2001 was replaced by 2002. The decade of the 1990s was the hottest at least in this last millennium. And the planet is heating at a rate faster than at any time in the last 10,000 years.

Senior scientist Tom M.L. Wigley of the National Center for Atmospheric Research put the coming changes in perspective in a letter to Senators Tom Daschle (D–ND) and William Frist (R–TN) in July 2003:

> There is only one chance in 100 that the rate of warming will be less than double the warming rate of the last 100 years—and a 99 percent probability that it will exceed double the past warming rate . . . The most likely estimate of warming between [now] and 2100 is 5.5 degrees F. This is five times the warming rate experienced over the past 100 years. At the high end, there is a five percent chance that the warming could be more than eight times the warming rate of the past century.

Our climate is capable of immense and wildly disruptive surprises. Every day, those surprises seem progressively more likely than not. Not only are we gambling with our future, we are gambling with our eyes blindfolded. We can't even read the cards we've been dealt.

SNAPSHOTS OF THE WARMING NO. 2

Although the disappearance of the glaciers is visually striking, its importance may be overshadowed by a less visible but more pervasive consequence of atmospheric warming. All over the world, species are traveling toward the poles in an effort to maintain temperature stability.

In early 2003, two major studies published in the journal Nature *found that the warming was accelerating the migration of species around the globe—even as it was significantly changing the timing of the seasons, especially in the Northern Hemisphere.*

One study showed that animals have shifted north an average of nearly four miles per decade. Another study showed that animals are migrating, hatching eggs, and bearing young an average of five days earlier than they did at the start of the twentieth century, when the average global temperature was 1°F cooler.

"There is a consistent signal. Animals and plants are being strongly affected by the warming of the globe," said Terry Root, a Stanford-based biologist and author of one of the studies. Root told the Los Angeles Times *she was surprised that the two*

studies were able to detect the effect. She said she thought the increased temperature was too small to cause widespread change. Root had expected that any damaging effects of climatic change would be unnoticeable amid the enormous habitat destruction in modern times caused by development, pollution, and other human activities.

"It was really quite a shock, given such a small temperature change," she said, adding: "If we're already seeing such dramatic changes among species, it's really pretty frightening to think what we might see in the next 100 years."

The studies noted, for example, that an Arctic seabird breeds twenty-four days earlier than it did decades ago and one species of butterfly shifted its range northward by nearly sixty miles over the last century.

A second study, co-authored by Camille Parmesan and Gary Yohe, found that some butterflies have shifted northward in Europe by thirty to sixty miles or more, with the changes closely matching those in average warm-season temperatures. Parmesan told the New York Times *that the researchers were able to rule out other factors—habitat destruction, for example—as causes of the changes. The Parmesan-Yohe study found that species' ranges were tending to shift toward the poles at some four miles a decade and that spring events, like egg laying or trees' flowering, were shifting to 2.3 days earlier per decade.*

The report came a year after a similar study was published by a team of German scientists. That study found that ecosystems around the globe are showing the effects of climate warming. Earlier arrival of migrant birds, earlier appearance of butterflies, earlier spawning in amphibians, earlier flowering of plants—spring has been coming sooner every year since the 1960s, researchers reported.

"Although we are only at an early stage in the projected trends of global warming, ecological responses to recent climate change are already clearly visible," wrote Gian-Reto Walther of the University of Hanover, Germany.

"There is now ample evidence that these recent climatic changes have affected a broad range of organisms with diverse geographical distributions," Walther and his team reported. "The implications of such large scale, consistent responses to relatively low average rates of climate change are large," the researchers warned, adding that "the projected warming for the coming decades raises even more concern about its ecological and socio-economic consequences."

In Britain, a recent study found in the fall of 2002 that British seasons are becoming increasingly muddled. Spring arrived three weeks early that year. Oak trees changed color more than a week late, while beech trees lost their leaves twelve days later than usual. People in certain parts of the United Kingdom found they had to mow their lawns all year-round.

What most startled officials was the rapidity with which the change is taking place. Dr. Tim Sparks of the Centre for Ecology and Hydrology said the most astonishing finding was how much earlier spring came in 2002 compared with 2001. "The majority of climate scientists would agree that there are already signs of a warming climate and this is having a knock-on effect on our plants and animals," he told the BBC.

The change in the timing of the seasons may seem of small consequence. On reflection, it is profoundly sobering. As the

writer Bill McKibben put it: "What this means is that, by acci-
dent, we are changing the rhythms of nature by which we have
lived our lives and planted our crops and written our poetry for
10,000 years."

As growing numbers of plants, animals, birds, insects, and
fish are becoming imperiled, their past and future habitats face
an equally bleak outlook.

On landmasses around the world, habitats are being rapidly
altered. According to one report, one-third of the world's habi-
tats could disappear or change beyond recognition by the end of
this century. In Russia, Canada, and Scandinavia, up to 70
percent of habitats could be lost. In the United States, much of
the spruce and fir forests of New England and New York state
may be wiped out if carbon dioxide emissions are not reduced,
according to the World Wide Fund for Nature. "This is not some
slow, controlled change we're talking about. It's fast, it's unpre-
dictable and it's unprecedented during human civilization,"
said Adam Markham, a co-author of the report. The study was
based on a "moderate" projection that concentrations of carbon
dioxide in the atmosphere will double from pre-industrial levels
during this century.

These forecasts were underscored at the beginning of 2004 by
a study in the journal Nature *by nineteen researchers from seven*
countries, warning that rising temperatures could doom more
than one-third of the planet's species to extinction by 2050.

3

Criminals Against Humanity

With all of the hysteria, all of the fear, all of the phony science, could it be that man-made global warming is the greatest hoax ever perpetrated on the American people? It sure sounds like it. ᴌ SENATOR JAMES INHOFE (R–OK) JULY 28, 2003

With its persistent denial of the climate crisis, the Bush administration has isolated the United States from its closest allies—and staked out a position that conflicts directly with the findings of more than 2,000 scientists from 100 countries in what is the largest and most rigorously peer-reviewed scientific collaboration in history.

Nothing has further alienated the United States from the rest of the world than the Bush administration's dismissal of global climate change. Even before the Iraq War, the president chose to break with America's long-standing international allies in favor of another set of domestic allies—his supporters in the coal and oil industries.

Today, the White House has become the East Coast branch office of ExxonMobil and Peabody coal, and climate change has become the preeminent case study of the contamination of our political system by money.

The diplomatic contrast is striking. There is virtually no debate in any other country in the world (with the possible exception of Russia) about what is happening to the climate. All the debates in the other countries are about policy—how to change national energy diets without wrecking national economies.

Holland, for example, recently completed a plan to cut its emissions by 80 percent in forty years. Germany has committed to cuts of 50 percent in fifty years. British prime minister Tony Blair has pledged the United Kingdom to reducing carbon emissions by 60 percent in fifty years. Even China, whose economy grew by 36 percent between 1995 and 2000, cut its emissions by 19 percent during the same period.

By contrast, the approach of the Bush administration is far more industry friendly. Bush's climate plan centers on a voluntary effort for industries to reduce the "greenhouse gas intensity" of their production technologies. In calling for a reduction of "greenhouse gas intensity" of 18 percent by 2012, the administration attempted to create the impression that it actually intended to reduce climate-altering carbon emissions. But even if the voluntary plan were met with total compliance, it would continue to allow emissions to rise, albeit at a marginally slower rate than they have been.

The bankruptcy of this approach was assured by the failure of several other voluntary plans put forth during the Clinton administration.

More damning, perhaps, was the conclusion of the U.S. General Accounting Office. Prior to the release of the Bush plan, carbon intensity had already been projected to drop by 14 percent by 2012 through expected increases in energy efficiency. Given that projection, the report concluded: "[T]he administration . . . did not provide a detailed rationale for the emissions intensity target that it established. We did not find a specific justification for the additional four-percentage-point reduction as opposed to any other target that could have been established—or what achieving a four percent reduction is specifically intended to accomplish. . . . It is therefore unclear to what extent the [plan] will contribute to the goal of reducing emissions and thus lowering emissions intensity by 2012."

The reason for this abdication of leadership is not hard to understand.

Nature's demand that humanity cut its use of coal and oil by 70 percent means the end of the fossil fuel industry as we know it.

No one is more aware of that fact than the top executives of the oil and coal industries and their hired guns, a tiny handful of industry-funded "greenhouse skeptics," most of whom are laughingstocks in the scientific community.

Since the early 1990s, the fossil fuel lobby has mounted an extremely effective campaign of deception and disinformation designed to persuade policymakers, the press, and the public that the issue of climate change is stuck in scientific uncertainty.

Since 1991, that campaign has attacked the science. It subsequently misrepresented the economics of a transition to clean energy. Most recently, with its new champion in the White House, it has tried to demolish the diplomatic foundations of the climate convention. And it has been extremely successful in maintaining a drumbeat of doubt in the public mind inside the United States.

With the 2000 presidential election, however, the fossil fuel lobby won a victory beyond its wildest dreams. What began as an industry campaign of deception and disinformation was adopted as presidential strategy.

Shortly before Bush's inauguration, at the end of 2000, climate activists thought they had scored a major victory when one of the most powerful industry groups opposing action on the climate fell apart. The Global Climate Coalition had represented more than 6 million businesses—including, among others, the American Highway Users Alliance, the American Petroleum Institute, the Edison Electric Institute, the National Association of Manufacturers, and the National Mining Association.

Under pressure from activists, British Petroleum and Shell left the coalition in 1999—fearful their continuing membership would permanently taint their public image. They were followed shortly thereafter by Ford, Daimler-Chrysler, Texaco, the Southern Company, and General Motors. As a result of these defections, the Global Climate Coalition was dissolved at the end of 2000.

The activists' victory, however, was short-lived. By 2001, many of the same industry associations and companies had a direct pipeline into the White House.

The official line of the new Bush presidency was expressed in a November 2002 memo from political consultant Frank Luntz to the Republican Party. In a section titled, "Winning the Global Warming Debate," Luntz wrote that many voters believe there is a lack of consensus about global warming among scientists. "Should the public come to believe that the scientific issues are settled, their views about global warming will change accordingly," he wrote. *"Therefore you need to continue to make the lack of scientific certainty a primary issue."*

The memo, obtained by the Environmental Working Group, was drafted by the same public relations firm that crafted Newt Gingrich's "Contract with America" in the mid-1990s. And it set the tone for the Bush administration's posture toward the climate crisis.

In the spring of 2001, the president reneged on his campaign promise to cap emissions from coal-powered plants.

He then announced his administration's energy plan, which is basically a fast track to climate chaos. It calls for the construction of 1,300 to 1,900 new power plants—most of them powered by coal, the most carbon-intensive of fuels.

Next, Bush withdrew the United States from the Kyoto negotiations, parroting the cynical line of the U.S. coal industry that the protocol is unfair to the United States because it exempts the developing countries from the first round of emissions cuts. He contended it was also harmful to the U.S. economy.

(Ironically, it was Bush's father, then president George H.W. Bush, who approved the exemption of the developing countries from the first round of cuts under the 1992 Rio Treaty. The reasoning behind that decision was that the countries of the North created the problem and have the resource to begin to deal with it. Once the industrial world had taken meaningful steps to address the climate crisis, the developing countries would then assume their own obligations under a subsequent round of the treaty.)

The president then dismissed the findings of the U.N. Intergovernmental Panel on Climate Change (IPCC) as an inconsequential "product of bureaucracy" when, in fact, it reflects the peer-reviewed findings of more than 2,000 scientists.

In a truly Orwellian stroke, the White House in the summer of 2003 altered a climate report by the U.S. Environmental Protection Agency (EPA), editing out all references to the dangerous impacts of climate change on the United States. The result was summed up smartly in a comment in *Harper's Index:*

> Number of paragraphs devoted to global warming in the EPA's 600 page Draft Report on the Environment of 2003: 1 paragraph.
>
> Number of paragraphs removed by the White House before publication: 5: Source: EPA.

Some apologists for the Bush administration have justified its climate policies by citing politically conservative principles—the withdrawal of onerous regulations, a belief in the unqualified efficacy of free markets, and the tepid appeal to corporate voluntarism.

In fact, the Bush climate policies have nothing to do with political conservatism. Rather, they represent corruption disguised as conservatism.

For starters, a number of officials in the Bush administration have close ties to the fossil fuel lobby. Vice President Dick Cheney was CEO of Halliburton, the country's largest oil field services firm. National Security Adviser Condoleezza Rice served on Chevron's board of directors. Secretary of Commerce Donald L. Evans worked for a Denver oil and gas company (Tom Brown, Inc.). Andrew Card, the White House chief of staff, is a former president of the American Automobile Manufacturers Association. Energy Secretary Spencer Abraham received over $700,000 in contributions for his 2000 Senate race from auto industry contributors. Interior Secretary Gale Norton, after receiving more than $285,000 from energy industries for her 1996 race in the Colorado state senate, then chaired the Coalition of Republican Environmental Advocates, which was funded, among others, by BP Amoco and Ford.

Bush's electoral success, moreover, was significantly aided by the support of big coal and big oil.

The result of the 2000 presidential race was due—in significant measure—to Bush's victory in West Virginia, a state no other Republican presidential candidate had ever won. (The pivotal nature of the West Virginia vote has been generally overshadowed by the controversy over the vote count in Florida.)

The Bush victory in West Virginia was a result of the substantial political and financial support of the state's coal industry. One coal executive alone, James Harless, raised $275,000 for the Bush campaign in West Virginia, which, according to the *Wall Street Journal*, was five times more than Al Gore raised in the state.

Overall, coal interests nationwide tripled their 1996 contributions and donated $3.8 million to 2000 presidential elections—88 percent of which went to the Republican Party. Electric utilities and their executives and employees gave $18.4 million to candidates and parties, of which $12.4 million went to Republicans, according to the Center for Responsive Politics.

Five months after Bush's inauguration, an official of the West Virginia Coal Association told a meeting of the organization: "You did everything you could to elect a Republican President. Now you are already seeing in his actions the payback . . . for what we did."

That "payback" came in the form of an about-face on a campaign promise that candidate Bush had made in 1999—to repeat what he had done as governor of Texas and impose a cap on carbon dioxide emissions from all coal-fired power plants in the state.

In early March 2001, Christine Todd Whitman, administrator of the Environmental Protection Agency, had announced that Bush would soon make good on his campaign promise to limit power plant emissions.

Two weeks later, the president pulled the rug out from under Whitman (the first of many such incidents) by reneging on that commitment. In a letter to four Republican senators, Bush declared he was backing away from the cap because of "incomplete state of scientific knowledge of the causes of, and solutions to, global climate change and the lack of commercially available technologies for removing and storing carbon dioxide."

One person who was particularly pleased by Bush's flip-flop was Irl Engelhardt, chairman of the Peabody Group, the country's biggest coal company. Engelhardt had donated $250,000 to the Republican National Committee—and served as an adviser to the Bush-Cheney Energy Transition Team.

The specific payback for Engelhardt's company, however, wouldn't come for several more months.

Shortly after taking office, Bush charged his vice president, Dick Cheney, with establishing a task force to craft a new national energy plan. On May 1, Cheney provided a teaser of that plan in a talk to the annual meeting of the Associated Press in Toronto.

Cheney declared that coal had been neglected in considering America's energy future. He noted his intention to address what he called the "most plentiful source of affordable energy" in the United States, adding that people who sought to phase out its use "deny reality." As for investments in energy conservation, Cheney noted: "Conservation may be a sign of personal virtue, but it is not a sufficient basis for a sound, comprehensive energy policy."

The work of Cheney's task force reached fruition two weeks later when, on May 17, he unveiled the first draft of the

administration's energy plan. The plan called for an expanded
role for nuclear power, the opening of the Arctic National
Wildlife Refuge for oil exploration, and the construction of between 1,300 and 1,900 new power plants—most of them powered by coal and nuclear energy—over the next twenty years.

One person who was clearly not thrilled with the plan
was Bush's then treasury secretary, Paul O'Neill, former
CEO of Alcoa. O'Neill had long lobbied for increased taxes
on gasoline and oil—and at one point had likened the coming impacts of climate change to a "nuclear holocaust." In
1998, O'Neill told the Aluminum Association that even
though real emission reductions were not mandated until the
year 2008, "I believe a real danger to civilization is that . . .
we don't do anything for 10 years."

If O'Neill, who later resigned, was profoundly disappointed with the administration's coal and oil-rich energy
plan, Engelhardt, the CEO of Peabody Energy, was not.

Peabody's management was extremely cozy with the
Cheney team. Its chief lobbyist, Fred Palmer, had, during the
1990s, headed the Western Fuels Association, a coal consortium that mounted relentless attacks on the findings of
mainstream climate scientists.

In 2001, shortly after the Bush election, Palmer and
other members of Peabody Energy met with the Cheney task
force a number of times while the administration's energy
plan was being fashioned. Peabody executives were among
thirty or forty industry officials briefed by members of the
task force in meetings set up by the Edison Electric Institute,
the power industry's primary lobbying group, according to a

report in the *New York Times.* Those same meetings were closed to the major environmental groups.

Shortly after the plan was announced, Peabody underwent a major corporate change.

For its 120-year life, Peabody, the country's biggest coal company, had been a privately held entity. But within a week after the Cheney plan was announced, Peabody issued an IPO and went public. Overnight, its stock jumped from $24 to $36. The IPO was handled by Lehman Brothers, a brokerage house that, itself, was the sixth-largest energy industry contributor to the Republican Party during the 1999–2000 election cycle.

If the administration's energy plan was covered with the fingerprints of the coal industry, its climate policies seem to have been directly dictated by the largest oil company on the planet.

On February 6, 2001, just weeks after Bush assumed office, an ExxonMobil official, Randy Randol, sent a memo to the White House that was subsequently obtained by the Natural Resources Defense Council through a series of Freedom of Information requests.

The memo cited a quote from Dr. Robert Watson, chairman of the IPCC, in which Watson noted: "The United States is way off meeting its targets. A country like China has done more, in my opinion, than a country like the United States to move forward in economic development while remaining environmentally sensitive."

It raised the question of Watson's tenure, under a section of the memo titled: "ISSUE: Can Watson Be Replaced Now at the Request of the United States?" ExxonMobil recommended that the Bush administration remove Watson, along with Rosina Bierbaum, an official in the Clinton White House Office of Science and Technology Policy, and Dr. Michael MacCracken, who headed up the U.S. Global Change Research Program—both of whom were instrumental in producing the "U.S. National Assessment on Climate Change," the EPA document that was eviscerated by the White House in the spring of 2003.

In their place, the oil giant recommended the appointment of several longtime climate "skeptics"—Dr. John Christy and Dr. Richard Lindzen. Christy, of the University of Alabama–Huntsville, had long cited satellite records to downplay the reality of climate change. Lindzen, an MIT climatologist, whose central thesis was recently disputed in *Scientific American*, has long dismissed climate change as a negligible or non-event.

The satellite argument fell apart in 1998 when two researchers discovered that the decay of the satellite orbits—on which Christy's thesis rested—had been calculated erroneously. When they were corrected, the satellites indicated a distinct warming trend. Lindzen, for his part, has acknowledged that he receives $2,500 a day to consult with coal interests in Australia and the United States and oil interests in the Organization of Petroleum Exporting Countries (OPEC). His own work, designed to indicate that atmosphere convection patterns and the role of clouds would render global warming a self-limiting event, was essentially refuted by researchers from NASA.

In 2001, Lindzen published research indicating that warming would decrease tropical cloud cover and therefore exert an overall cooling impact on the earth's surface. But three research teams arrived at different conclusions. One team of NASA researchers, for example, found that Lindzen used satellite images of tropical clouds that were not representative of overall tropical cloud conditions. The team, headed by Bruce Wielicki of the NASA Langley Research Center, concluded that the tropical clouds would have a slight heating effect, according to a report in *Scientific American*.

Ultimately, the Bush administration did succeed in scuttling Watson's chairmanship of the IPCC. But, apparently fearing a major backlash, the administration decided not to back such visible contrarians as Christy or Lindzen for the position. Instead, it supported the nomination of Dr. Rajendra Pachauri, an Indian economist with an engineering background. Pachauri, who had headed the Tata Energy Research Institute in India, was known to have worked closely with the electric utility industry. But despite his energy background, Pachauri has come to embrace the science—and has been outspoken about the vulnerability of developing countries to climate impacts.

However, it was in the area of diplomacy rather than science that ExxonMobil most strongly directed the policies of the Bush administration.

ExxonMobil officials also recommended that President Bush hire Harlan Watson (no relation to Robert), a Republican staffer on the House Science Committee, to work on the new Bush climate team. Not long afterward, the White House announced that Watson had been appointed the administration's chief climate negotiator.

His position was clear from the outset. In May 2002, Harlan Watson declared that the United States would not engage the Kyoto process for at least ten years. In a talk in London, Watson announced the White House "wanted no part" of a 2005 international review of greenhouse gas reductions. "The next time we take stock on climate change," he said, "has been set by the president at 2012."

In an effort to improve its environmental image, Exxon-Mobil announced in November 2002 that it was investing in research into hydrogen fuel. The company noted it intends to derive hydrogen from oil and coal—a process that creates so many carbon emissions as to virtually neutralize the climate-friendly benefits of the fuel. (Hydrogen can be produced just as easily and with far less negative environmental impact if it is derived from water rather than from coal or oil. If the electricity for the process is produced from renewable sources, it creates virtually no carbon emissions.)

ExxonMobil also announced it was investing in research into a hugely expensive—and completely untested—system of mechanical carbon sequestration in which massive amounts of carbon dioxide would be piped into deep mountain wells.

Two months later in his State of the Union address, Bush announced his own petroleum-based hydrogen initiative that directly reflected the approach of ExxonMobil in preserving America's oil infrastructure.

ExxonMobil's new public relations charade did little to conceal its real intentions. By 2001, ExxonMobil had replaced the coal industry as the major funder of the most prominent and visible "greenhouse skeptics." By 2003, ExxonMobil was giving more than $1 million a year to an array of ideological, right-wing organizations opposing action on climate change—including the Competitive Enterprise Institute, Frontiers of Freedom, the George C. Marshall Institute, the American Council for Capital Formation Center for Policy Research, and the American Legislative Exchange Council.

In its effort to sabotage the unprecedented scientific consensus of the IPCC, ExxonMobil has basically picked up where the coal industry left off.

During the 1990s, that effort had been spearheaded by Fred Palmer, who, around the time of the Bush election, was hired as chief lobbyist for Peabody Energy. Prior to his hiring by Peabody, Palmer headed up the Western Fuels Association, a $400-million coal consortium that had funded a tiny handful of industry-funded "greenhouse skeptics" who had long been dismissed by the mainstream scientific community.

Throughout the 1990s, Palmer directed an extensive and extremely successful public relations offensive funded by the coal industry that used such prominent "greenhouse skeptics" as Fred Singer, Pat Michaels, Sherwood Idso, and Robert Balling, among others. One campaign, which sent three of these "skeptics" around the country to do media interviews, was crafted, according to its strategy papers, "to reposition global warming as theory rather than fact" and, more specifically, was designed to target "older, less-educated men . . . and young low-income women" in districts that get their electricity from coal and preferably have a member on the House Energy Committee, according to the strategy papers for the campaign.

Over the last ten years, those skeptics received more than a million dollars, either directly or indirectly, from coal and oil interests. Their strategy was quite simple—continue to raise doubts about the science in order to preempt any public demand for action. Their funding by the fossil fuel lobby was never disclosed publicly until it was published in *The Heat Is On* in 1997. (The issue of financial disclosure is not a small one. Industry-funded research can be neutral—and it can be good or bad. But disclosure is critical so that the work in question can be reviewed with an eye to commercial bias. If, for instance, a medical researcher's work is funded by a pharmaceutical company, that funding must be declared in the tag line as a condition of publication. Unfortunately, those same guidelines do not apply to climate science. And—most damning—few journalists who have written about this issue have ever bothered to ask about funding.)

What is especially telling about the industry-funded "greenhouse skeptics" is their lack of standing in the scientific community. In a review of Michaels's work, Tom M.L. Wigley, a preeminent climate modeler at the National Center for Atmospheric Research, concluded it was so flawed that not only would it fail to pass the scrutiny of qualified climate scientists, it would not even be accepted for peer review.

As for Singer, he was unable to publish in the peer-reviewed literature for twenty years (save for one technical comment) until mid 2004, when he co-authored two refereed papers.

Singer's recklessness transcends his deeply flawed scientific pronouncements. It involves at least one public lie about his own funding. In early 2001, Singer was accused of having his work funded by the oil industry. In response, Singer wrote in a letter to the *Washington Post* that he had not received any oil industry money for at least twenty years—when he had done a consulting job for the industry.

In fact, Singer received at least $10,000 and as much as $75,000 from ExxonMobil in 1998 alone, according to information on the oil giant's own Web site. (Shortly after that information was published in the *Nation*, ExxonMobil withdrew the page from its Web site.)

Overall, however, the success of the campaign of disinformation by the fossil fuel lobby on the public and on policymakers in the United States is striking. One proof of the success of that campaign is reflected by two polls done by *Newsweek* magazine. Back in 1991, 35 percent of people surveyed by *Newsweek* said they thought global warming was a very serious problem. By 1996—even though the science had become far more robust and the IPCC had declared that humans are, indeed, changing the climate—that 35 percent had shrunk to 22 percent, largely as a result of the fossil fuel lobby's deceptive public relations campaign.

That record of success was clearly not lost on the Luntz group.

One section of the notorious Luntz memo counsels the president: "The most important principle in any discussion of global warming is your commitment to sound science. Americans unanimously believe all environmental rules and regulations should be based on sound science and common sense. Similarly our confidence in the ability of science and technology to solve our nation's ills is second to none. Both perceptions will work in your favor if properly cultivated."

Sure enough, the most prominent new effort by the skeptics to discredit the findings of mainstream scientists surfaced in the spring of 2003 in the form of a study authored by Sallie Baliunas and Willie Soon at the Harvard-Smithsonian Center for Astrophysics and published in an obscure journal, *Climate Research*. The paper was coauthored by Craig Idso and Sherwood Idso, whose Center for the Study of Carbon Dioxide was long funded by the coal industry and more recently supported by ExxonMobil.

The study was immediately picked up by several congressional Republicans as a shining example of "sound science."

In their review of the literature, Baliunas and Soon concluded that the twentieth century is neither the warmest century nor the century with the most extreme weather of the past 1,000 years. Both researchers had previously contended that the recent extreme and rapid warming was due, almost entirely, to solar variations—a finding that had long since been disproved by a number of peer-reviewed scientific studies.

The recent report by Baliunas and Soon was further undermined when it was revealed that their work was funded by, among others, the American Petroleum Institute.

Nevertheless, the study was seized upon by Senator James Inhofe (R–OK), chair of the Senate Environment and Public Works Committee, who said the science showed that natural variability, not human activity, was the "overwhelming factor" influencing climate change. It is probably not a coincidence that Inhofe received nearly twice as much in campaign contributions from energy companies during the 2002 election as from any other business sector.

In short order, a lead author of the IPCC's Third Assessment Report, Dr. Michael Mann, branded the Baliunas-Soon study "nonsense" and said the science was "fundamentally unsound." That was followed by a letter, signed by thirteen leading climate scientists from the United States and the United Kingdom, including Mann, to the peer-reviewed journal *Eos*, of the American Geophysical Union, which declared that the anomalous late-twentieth-century warmth cannot be explained without taking into account the contributions of human activities.

(It was Mann who coauthored a recent research paper with Phil Jones, director of climate research at the University of East Anglia in the United Kingdom, indicating that the planet is hotter today than at any time in the last 2,000 years. The paper was published in the refereed journal *Geophysical Research Letters* in August 2003.)

Most telling, perhaps, was another piece of fallout from the Soon-Baliunas controversy that received little news coverage.

Three editors of the journal that published the skeptics' study resigned after they were forbidden from writing an editorial pointing out the methodological errors by the industry-funded researchers.

"They submitted a flawed paper," said Hans von Storch, editor in chief of the journal *Climate Research*. He added that the owners of the magazine, which is published in Germany, refused to allow him to write an editorial saying the paper was flawed. As a result, von Storch and two other editors at the journal resigned in protest over the Soon-Baliunas work.

While the Bush administration was attempting to disguise its policies under the cloak of "sound science," the White House was working to scuttle the most thoroughly scientifically vetted projections of coming climate impacts inside the United States.

The report, known as the "U.S. National Assessment of the Potential Consequences of Climate Variability and Change" (USNA), is a meticulous document that was thoroughly peer reviewed and drafted under the auspices of Dr. Michael MacCracken, head of the U.S. Global Change Research Program during the Clinton administration. The national assessment is best described as an IPCC report on climate impacts—with an exclusively domestic focus. It carefully details a range of anticipated impacts in the various geographical regions and sectors (forests, grasslands, coastlines, etc.) of the United States.

The attack on the document from the fossil fuel lobby and its ideological allies surfaced in early 2002 following the passage of an obscure law known as the Federal Data Quality

Act, which requires that information disseminated by the government be useful, reliable, and reproducible. The Center for Regulatory Effectiveness, a politically conservative think tank, wrote to the Office of Management and Budget, as well as to the Office of Science and Technology Policy, and invoked the Data Quality Act, demanding that the assessment be withdrawn because it failed to meet standards of scientific objectivity. Not surprisingly, the head of the center, James Tozzi, was a longtime lobbyist for Philip Morris, General Motors, and groups representing the chemical and forestry industries, among others. The critique of the national assessment, which purported to point out its scientific flaws, was written by Pat Michaels, a climatologist from Virginia who has been on the coal industry's payroll since the early 1990s.

When that avenue failed to produce results, the politically conservative Competitive Enterprise Institute (CEI) sued the White House's environmental arm, the Council on Environmental Quality, to remove the document from circulation. In its press release, the CEI stated, "These junk science reports are already being used . . . by global warming alarmists and states seeking to hobble those more competitive."

In August 2003, the attorneys general of Maine and Connecticut made an extraordinary discovery. Through a Freedom of Information request, they unearthed e-mails indicating that the White House had secretly requested the private, right-wing CEI to sue it—the White House—in order to have the national assessment withdrawn.

The attorneys general cited a communication between the White House and the Competitive Enterprise Institute in which Philip Cooney, then chief of staff at the White House Council on Environmental Quality, asked the CEI to initiate a suit against the White House to have the document removed. Prior to his appointment at the White House, Cooney had headed up the climate program for the American Petroleum Institute.

"It appears that certain White House officials conspired with an anti-environmental special interest group to cause the lawsuit to be filed against the federal government," said Maine attorney general Steven Rowe, who complained to the U.S. Department of Justice along with Connecticut attorney general Richard Blumenthal.

Nor is it a coincidence that the Competitive Enterprise Institute, which brought the suit, is funded in part by ExxonMobil.

Perhaps the most seductive argument of the greenhouse skeptics and their sponsors in the coal and oil industries is that warming will be beneficial in certain areas. Having been forced by the weight of scientific findings to abandon their assertion that climate change is not happening, the fossil fuel lobby now contends it will be good for us. Their spokesmen now assert that a warmer climate in the United States will lower heating bills, expand agricultural areas, and reduce cold-weather mortality rates. The problem with this argument is that it fails to acknowledge the dynamic nature of climate change. We cannot arrest global warming at any particular comfort level. Once large planetary systems become destabilized, their momentum does not stop until they settle into a new—and potentially very different—state of equilibrium. Embedded in the climate system are large and power-ful feedback mechanisms that feed off each other and, be-yond a certain point, trigger a cascade of irreversible changes.

For one example, the tundra in northern Canada has ab-sorbed large quantities of heat-trapping methane and carbon dioxide. As the temperature rises, the tundra is beginning to thaw and release those gases back into the atmosphere. As that process proceeds, it will likely trigger a rapid spurt of new warming that, in turn, will increase the water require-ments of heat-stressed food crops while at the same time ac-celerating the decline in available water supplies. Similarly, the warming of the Arctic Ocean has thinned the ice cover so dramatically that freighters can now sail across the top of the North American continent in open water. But that boon to shipping could be a prelude to catastrophe. As the Arctic ice continues to melt, it is diluting the salt content of the North Atlantic. That change in salinity, in turn, could slow or stop the movement of a major Atlantic Ocean current whose warmth has kept the climate of northern North America and northern Europe hospitable to civilization. Should that cur-rent fail due to the melting of the Arctic ice cover, the result would be a rapid deep-freeze that would basically make much of the Northern Hemisphere incapable of supporting complex, urban societies.

There may be some short-term benefits from global warming. In the long run, they will be no more than a pleas-ant interlude before an indeterminate nightmare.

Stepping back, it is clear that climate change represents a titanic clash of interests that pits the ability of the planet to support this civilization against the survival as we know it of one of the biggest commercial enterprises in history—big coal and big oil.

Nowhere is the profound division between the oil-dominated Bush administration and the rest of the world more visible than in the area of climate change.

As of September 2003, 116 countries had signed or ratified the Kyoto Protocol—indicating a broad base of support despite the protocol's potential demise at the hands of Russia. Several industrial nations were committing themselves to carbon cuts that were an order of magnitude greater than the protocol's initial requirement of reducing national carbon emissions to 1990 levels by 2012. Moreover, a substantial number of developing countries—which are not required to cut their emissions in the first round of the protocol—have already begun to do so.

Clearly, these governments would not be undertaking these massive and wrenching changes if there were any doubt about what is happening to the planet.

When Bush withdrew the United States from the Kyoto Protocol in 2001, he triggered a profound swell of outrage throughout the rest of the world. That outrage was temporarily dampened by the need to forge an international anti-terror coalition in the wake of the attacks of September 11, 2001.

But it has not gone away.

Given the momentum of climatic instability—and the willingness of the rest of the world to begin to address it—the rage against the U.S. indifference to humanity's common future still simmers just below the surface.

In the early 1990s, when the science was still uncertain, this denial by the fossil fuel lobby could be excused as a predictable, business-as-usual response. But since the science has become so robust and the impacts so visible, this behavior now constitutes a clear crime against humanity.

The industry-funded campaign goes far beyond traditional public relations spin. It basically amounts to the privatization of truth.

Not that the corruption of our political system by the fossil fuel industry is that much different from other instances of corruption that we have, unfortunately, come to accept.

The difference is the context—and the stakes that are involved this time around. This time the corruption is not leading to some unemployed workers, some defective products, or some diminished pension funds. This time it involves the future of this civilization.

Our fossil fuels have brought us to a level of abundance and prosperity that was unimaginable a century ago.

Today they are propelling us forward into a century of disintegration.

Snapshots of the Warming No. 3

With water covering about two-thirds of the planet, the warming of the world's oceans—down to depths of nearly two miles—is progressing apace with the warming of the atmosphere. The warming of surface waters is exerting profound impacts on marine populations, coral reefs, and whole countries.

In 2002, scientists announced findings in Portugal that echoed similar findings in Monterey Bay, off the California coast, seven years earlier. Rising water temperatures are dramatically changing the composition of marine populations. In Portugal's Tejo River estuary, the biggest in Western Europe, global warming has caused such cold-water species as flounder and red mullet almost to disappear in the last two decades, according to Maria José Costa, director of oceanography at the University of Lisbon. At the same time, the numbers of warm-water fish such as Senegal sea bream, common to North African waters, and dogfish have vastly increased.

"This can only be explained by the temperature going up," said Costa.

One consequence of the warming of the world's oceans is a strong—and negative—impact on all types of marine life. A report by the World Wildlife Fund and Marine Conservation Biology Institute found rising global temperatures impacting ocean ecosystems to a far greater extent than previously acknowledged. From the tropics to the poles, widespread changes in marine life are occurring in step with rising water temperatures. The newly assembled evidence shows dramatic impacts arriving sooner than predicted.

Among the report's most disturbing news is research suggesting that Pacific salmon may no longer find suitable habitat in the Pacific Ocean. Other effects of the warming climate are appearing across the marine food chain, from plankton, penguins, and polar bears to fisheries on which humans depend.

Most dramatic of all is the sheer scope of the data.

"Warmer temperatures are raising the biological cost of living for marine species," said Dr. Elliot Norse, president of the Marine Conservation Biology Institute. "This is true in polar seas, where climate changes have been most pronounced, as well as on tropical coral reefs, which are suffering unprecedented devastation due to heat stress."

The shrinkage of most of the world's coral reefs from rising ocean temperatures was documented in 2001 by the United Nations Environment Programme (UNEP). That study of the "rainforests of the oceans" showed that 58 percent of the world's coral reefs are threatened by human activities—and that certain reefs have shrunk by as much as 90 percent due to ocean warming and other assaults. The coral reefs "are rapidly being degraded by human activities. They are over-fished, bombed and poisoned," said UNEP director Klaus Topfer. "They are damaged by irresponsible tourism and are being severely stressed by the warming of the world's oceans. Each of these pressures is bad enough in itself, but together, the cocktail is proving lethal." Although the damage to coral reefs is taking place in every part of the world, the worst damage was found to have occurred in the Indian Ocean, where warming oceans contributed to the loss of about 90 percent of certain coral reefs.

The ocean warming has also triggered the emergence—or resurgence—of a host of diseases affecting marine life. Warming waters have nurtured disease organisms that have accounted for, among other things, the extinction of a species of sea urchins in the Caribbean, the destruction of corals and sea grasses, an increase in blooms of toxic algae, the spread of a form of distemper among seals in Antarctica and Lake Baikal, and viral disease outbreaks among porpoises and dolphins.

In a startling report in late 2003, researchers reported that the entire ecosystem of the North Sea is in a state of collapse—with devastating implications for fisheries and wildlife. According to the report, "record sea temperatures are killing off the plankton on which all life in the sea depends, because they underpin the entire marine food chain. Fish stocks and sea bird populations have slumped.

"Scientists at the Sir Alistair Hardy Foundation for Ocean Science in Plymouth, which has been monitoring plankton in the North Sea for over 70 years, say that an unprecedented heating of the waters has driven the cold-water species of this *microscopic but vital food hundreds of miles to the north. They have been replaced by smaller, warm-water species that are less nutritious."*

"A regime shift has taken place and the whole ecology of the North Sea has changed quite dramatically," Dr. Chris Reid, di-*rector of the foundation, told the* Independent, *adding: "We are seeing a collapse in the system as we knew it.*

4

Bad Press

[T]he press's adherence to balance actually leads to biased coverage of global warming . . . This bias, hidden behind the veil of journalistic balance, creates . . . political space for the U.S. government to shirk responsibility and delay action regarding global warming.

᷂ MAXWELL T. BOYKOFF AND JULES M. BOYKOFF, "BALANCE AS BIAS: GLOBAL WARMING AND THE U.S. PRESTIGE PRESS," *Global Environmental Change* 14 (1)
JUNE 2004

If the public relations specialists of the oil and coal industries are criminals against humanity, the U.S. press has basically played the role of unwitting accomplice by consistently minimizing this story, if not burying it from public view altogether.

In 1997, Bert Bolin, a Swedish meteorologist who was, at the time, chairman of the Intergovernmental Panel on Climate Change, declared: "The large majority of governments, while recognizing uncertainties, believe that we know enough to take action now. This position was supported by an independent group of 2,000 scientists."

Or, as James McCarthy, who would later chair Working Group II of the IPCC, noted several years ago: "There is no debate among any statured scientists working on this issue about the larger trends of what is happening to the climate."

That is something you would never know from the press coverage.

Although the scientific community has known since 1995 that we are changing our climate, the U.S. press has done a deplorable job in disseminating that information. The singular exception lies in the reporting of Andrew Revkin, who covers climate change for the *New York Times*.

There are a number of reasons for this—none of them, given the magnitude of the story, justifiable.

On a somewhat superficial level, the career path to the top at news outlets normally lies in following the track of political reporting. Top editors tend to see all issues through a political lens.

For instance, although climate change has been the focus of a number of feature stories (and small, normally buried reports of scientific findings), the only time it has gained real news prominence is when it has played a role in the country's politics. During the 1988 presidential campaign, the first President Bush slapped the label of "ozone man" on Al Gore because of his book, *Earth in the Balance*. (It does not seem to be a coincidence that Gore totally ran away from the climate issue during the 2000 campaign.)

The issue again received prominent coverage in 1997 when the Senate voted overwhelmingly not to ratify the Kyoto Protocol—not because of the substance but because it signaled a political setback for the Clinton administration at the hands of congressional Republicans. Remarkably, the press paid scant attention to an industry-funded advertising blitz in the run-up to that vote. That campaign, which cost $13 million, centered on the message that the Kyoto Protocol "isn't global and it isn't fair" (because it exempts the developing countries from the first round of emissions reductions). Tellingly, the ads all appeared in Washington- and New York-based media outlets that were read by the real targets of the campaign—U.S. senators.

Most recently, the issue surfaced when President Bush withdrew the United States from the Kyoto process. Again, the coverage focused not on climate change but on resulting diplomatic tensions between the United States and the European Union (EU).

Prior to his withdrawal from Kyoto, President Bush declared he would not accept the findings of the IPCC—because the organization represented "foreign science," even though about half of the 2,000 scientists whose work contributes to the IPCC reports are American. Instead, Bush called for a report from the U.S. National Academy of Sciences that would provide "American science." The subsequent response from the NAS not only affirmed the findings of the IPCC but indicated that the IPCC may have even understated the magnitude of some coming impacts.

Astonishingly, even as the Washington press corps reported this story, few—if any—reporters bothered to check the position of the NAS. Had they done so, they would have found that as early as 1992, three years before the IPCC determined that we are changing the climate by our burning of oil and coal, the NAS recommended strong measures to minimize climate impacts.

The culture of journalism is, basically, a political culture that is not particularly hospitable—that is, in fact, institutionally arrogant—toward nonpolitical areas of coverage.

If the press were disposed to look beyond just the politics of Kyoto, it would be an eye-opener for the American public.

Aside from the pledges by Holland, Germany, and Britain to cut emissions by 50 to 80 percent in the next half century, the efforts by other countries to begin to address the climate crisis stand in vivid contrast to the indifference of the United States.

That contrast is apparent in the difference between the coverage of the climate crisis in the American press and the news media in other countries. While there has been no systematic and thorough analysis of comparative media coverage of the climate crisis in different countries, one recent study compared the attention given to the climate by the *Washington Post*, the *New York Times,* and the *Los Angeles Times* to three major newspapers in Britain and Germany. According to a weighted sampling between September 1999 and March 2000, the coverage in Britain was almost twice that of the press in the United States. The British paper, the *Guardian*, for example, accorded more than three times more coverage to the climate issue than the *Washington Post,* more than twice the coverage of the *New York Times*, and nearly five times more coverage than the *Los Angeles Times*. The German papers surveyed during the same period provided more coverage than the U.S. press—but less than expected, given the prominence of climate and energy issues in Germany's political life. Anja Kollmuss, who conducted the study, attributed that result to the fact that her sample spanned a period in which the German press was in full pursuit of a major financial scandal involving former prime minister Helmut Kohl.

In June 2003, the European Union agreed on a compact to reduce carbon fuel use through a system of "emissions trading" that will take effect in 2005. The EU pledged to cut emissions by 8 percent below 1990 levels, by 2010. In December 2002, the fifteen EU governments established a system in which companies in industries that are especially energy intensive will be assigned quotas for carbon dioxide emissions. Those who exceed their limit will be able to buy extra quota from others that stay below their allotted levels. The trading system will cover emissions from the power and heating industries, together with producers of steel, cement, glass, tile, paper, and cardboard. The story was prominently featured in the European press but was virtually ignored in the United States.

Nor have American journalists paid much attention to the growth of renewable energy around the world. Wind power in Europe, as one example, has been growing at a rate of 40 percent a year—much of it in the form of offshore wind farms. "It's going so fast now because there is a race to go offshore, with manufacturers and utilities competing for the jobs," said Corin Millais of the European Wind Energy Association.

"Companies are now talking of wind fields, like oil reserves or coal reserves, waiting to be tapped," Millais added.

Journalists might also have done a bit of checking on President Bush's assertion that one reason the United States has refused to accept emission reduction goals is because it would put the nation at a competitive disadvantage relative to developing countries.

In fact, many developing countries have taken very significant strides in this area. Through its development of hydropower and natural gas, for instance, Argentina has cut emissions by about 500 million tons over a twenty-five-year period. India is developing and deploying a range of climate-friendly technologies, including solar-electric facilities in rural areas, fuel cells for transportation, an array of wind farms, and the use of biomass to generate electricity. Even China, with its vast deposits of coal, managed to cut its greenhouse emissions by 19 percent during a five-year period in which its economy grew by 36 percent.

Were journalists to look beyond short-term political implications, their reporting would bring home how profoundly out of step the United States is relative to the rest of the world.

The next reason the issue is so neglected by the U.S. media has to do with the campaign of disinformation perpetrated by big coal and big oil. Although that campaign targeted the public and the policymakers, it also had a profound effect on journalists.

For many years, the press accorded the same weight to the "skeptics" as it did to mainstream scientists. This was done in the name of journalistic balance. In fact, it was journalistic laziness.

The ethic of journalistic balance comes into play when there is a story involving opinion: Should abortion be legal? Should we invade Iraq? Should we have bilingual education or English immersion? At that point, an ethical journalist is obligated to give each competing view its most articulate presentation—and equivalent space.

But when it's a question of fact, it's up to a reporter to dig into a story and find out what the facts are. The issue of balance is not relevant when the focus of a story is factual. In this case, what is known about the climate comes from the largest and most rigorously peer-reviewed scientific collaboration in history.

As James Baker, head of the U.S. National Atmospheric and Oceanic Administration, said, "There's no better scientific consensus on any other issue I know—except perhaps Newton's second law of dynamics."

Granted, there may be a few credentialed scientists—most notably Richard Lindzen—who have published in the peer-reviewed literature and who minimize climate change as relatively inconsequential.

In that case, if balance is required, it would suggest that a reporter spend a little time reviewing the literature, talking to some scientists on background, learning where the weight of scientific opinion lay—and reflecting *that* balance in his or her reporting. That kind of truly accurate balance would have reflected the position of mainstream scientists in 95 percent of the story—with the skeptics getting a paragraph at the end.

Today, that is finally beginning to happen.

A separate explanation for the failure of journalists to cover the climate crisis thoroughly lies in the fact that few journalists are comfortable with complex scientific information. Although a small number of news outlets have permanent science or environmental reporters on their staffs, more typically scientific and environmental stories are covered by general assignment reporters with no background in complex, scientific data. That lack of preparation is compounded by the daily deadlines that frequently deprive reporters of the time to fully digest complex scientific papers.

In fairness, the problem is compounded by many scientists. In their public statements, most scientists use extremely conservative and qualified language. Although this circumspect language is a requirement of approved scientific discourse, it leaves many journalists uncertain as to how meaningful a particular finding is.

One way to cut through this problem is through the time-honored use of background conversations with scientists. On the record, scientists typically speak in terms of probabilities and estimates and uncertainties. As a result, they sound to an untrained reporter as vague, wishy-washy, almost indecisive. But off the record, when asked to distill the implications of their findings, many scientists would make such statements as, "This is scary as hell." For a journalist who is not equipped to assess the relevance of a new computer model study, for example, the best fallback is to discuss the finding with scientists on background—and to solicit informal assessments from other scientists who specialize in the same area. Although background conversations do not provide quotes, they are essential to a reporter's understanding of the finding itself and enhance the ability of the reporter to put that information in a useful, interpretive context.

Background discussions can be extremely helpful in assessing the dimensions and reporting requirements of stories without compromising the identity of sources.

Remarkably, many journalists shy away from the primary source of climate science information—articles in the peer-reviewed literature. Most scientists write very clearly and economically. These papers, while frequently understated, are not beyond the comprehension of journalists. For a lay reporter, virtually the only science papers that may be beyond comprehension are those that center on computer models and involve extremely high-level mathematics. In those cases, background discussions with the researchers are necessary, unless the reporter has been educated in the area of computer modeling. But the vast majority of the scientific papers on climate change are quite accessible if one is willing to take the time to read them.

At a conference of the Society of Environmental Journalists several years ago, one veteran reporter from a large newspaper confessed that he had recently read a scientific paper for the first time, rather than relying on the summaries of others. He characterized it as a liberating experience to read the literature firsthand. Many reporters in the room responded as though this were a revelation—rather than an embarrassing acknowledgment of journalistic laziness.

Today climate change is no longer an issue of atmospheric science, although many scientific uncertainties remain—for instance, the role of clouds, future rates of warming, and specific impacts in particular geographic areas, to name a few. But the overwhelming predominance of climate research today focuses on the impacts of that warming. And those impacts are not beyond the grasp of journalists.

Any reporter who really wanted to make climate change more accessible to a general audience would need to look no further than the weather reports.

One of the first signs of early-stage global warming is an increase in weather extremes—longer droughts, more heat waves, more severe storms, and much more intense, severe dumps of rain and snow. Today, extreme weather events constitute a much larger portion of news budgets than they did twenty years ago.

Global warming, even without the amplification of periodic El Niños—is palpably changing the nature of our weather. It is almost as though nature is saying: "Look out the window. Time's up."

Following up on an earlier landmark study by Tom Karl, David Easterling of the National Climatic Data Center reported in a September 2000 article in *Science* that as the atmosphere warms, droughts, floods, heat waves, heavy rainfall, tropical storms, and hurricanes are expected to increase.

Wrote Easterling: "Our review shows consistency between our climate models and what we have observed in the 20th century. Models of 21st-century climate suggest that many of these changes in climate extremes are likely to continue."

Those findings were underscored by a groundbreaking report released by the World Meteorological Organization in July 2003. As the British newspaper the *Independent* reported: "In an astonishing announcement on global warming and extreme weather, the World Meteorological Organization signaled that the world's weather is going haywire.

"The WMO concluded that these record extreme events (high temperatures, low temperatures and high rainfall amounts, and droughts) have been gradually increasing over the past 100 years. New record extreme events occur every year somewhere in the globe, but in recent years the number of such extremes have been increasing."

The physics behind the altered drought and rainfall patterns are not extraordinarily complicated: As the atmosphere warms, it accelerates the evaporation of surface waters. It also warms the ocean waters, speeding up their evaporation rates. The heated air expands to hold more water. When the normal turbulence comes through the atmosphere, it results in much more intense downpours. The warming air also redistributes the moisture within the atmosphere—leading to more intense storms and rainfalls and more prolonged and protracted droughts.

The destructive power of more intense downpours was highlighted in a report by ABC News in which Peggy LeMone, a senior scientist at the National Center for Atmospheric Research, was asked to calculate the weight of a small, white cumulus cloud. "The water in the little cloud weighs about 550 tons," she said. "Or if you want to convert it to something that might be a little more meaningful . . . think of elephants." Since a normal adult elephant weighs about six tons, she said, that would mean that water inside one typical cumulus cloud would weigh about 100 elephants.

Many scientists believe we have already crossed into a new weather regime marked by extremes of all kinds.

Take the year 2001 as one example.

At the beginning of the year, Britain emerged from its wettest winter in more than 270 years of record keeping. In January and February, twenty-two successive blizzards in northern China stranded more than 100,000 herders, many of whom starved. In South Florida, the worst drought in 100 years decimated citrus crops, prompted extensive water restrictions, and triggered the spread of more than 1,200 wildfires. In early May, some forty people died in the hottest spring on record in Pakistan. In June, Houston suffered the single most expensive storm in modern history when tropical storm Allison dropped thirty-five inches of rain in one week, leaving $6 billion in damages. By late July, a protracted drought in Central America had left more than 1.5 million farmers with no crops to harvest—and 1 million people verging on malnutrition. In Iran, a devastating drought left more than $2.5 billion in agricultural losses. (The drought was temporarily interrupted in August by Iran's worst flash flooding in 200 years, which killed nearly 500 people.) In October, meteorologists documented a record ninety-two tornadoes in what is normally a quiet period for these events. In November, the worst flooding in memory killed more than 1,000 people in Algeria. In Boston, after an October and November of record-setting warmth, it was 71°F on December 1.

In the following year, 2002, more than 1,000 people died from a spring heat wave in India. The summer's floods in Russia, the Czech Republic, and Germany were the worst in memory. Wildfires consumed more than 5 million acres in the western United States and northern Canada. Drought conditions spread over half the United States. Back in India, 235 million people were plunged into darkness when the electricity grid collapsed because its hydroelectric sources dried up. Health officials reported locally transmitted cases of malaria in northern Virginia. West Nile virus spread to forty-two states—and, even more disturbing, to more than 230 species of mammals, insects, and birds. (Stagnant pools from downpours, which follow extended dry periods, create fertile breeding ground for mosquitoes that spread malaria and West Nile virus.) In South Asia, more than 12 million people were displaced by severe flooding.

In the spring of 2003, 1,400 people died from a heat wave in India and Pakistan. The United States experienced a record 562 tornadoes in the month of May. A brutal heat wave in Europe set new temperature records in Britain, triggered Portugal's worst forest fires in fifty years, and killed as many as 11,000 people in France in a four-week period.

Given the dramatic increase in extreme weather events, one might think that journalists, in covering these stories, would include the line: "Scientists associate this pattern of violent weather with global warming." They don't.

A few years ago, a top editor at a major TV network was asked why, given the increasing proportion of news budgets dedicated to weather disasters, the network news broadcasts did not make this connection. The editor said, "We did that. Once. But it triggered a barrage of complaints from the Global Climate Coalition to our top executives at the network." (The GCC was, at the time, the main fossil fuel industry lobbying group opposing action on global warming.)

The lobbyists subtly changed the subject by arguing that you can't attribute any one extreme event to climate change—just as you cannot attribute any one case of lung cancer to smoking. But that is off point. The scientific community is unambiguous in its finding that the first and most visible manifestation of the planet's warming is an increase in violent weather extremes.

The editor agreed that it would be very useful to the public in covering severe floods, droughts, and storms to note that "scientists associated this *pattern* of violent weather with global warming." But in the end, he confided, the industry basically intimidated the network into dropping this connection from its coverage. The threat was implicit: If the network persisted, it ran the risk of losing a lot of lucrative oil and auto advertising dollars.

Beyond the connection with extreme weather events lies a deeper betrayal of trust here by the media. By now, most reporters and editors have heard enough from environmentalists to know that global warming could, at least, have potentially catastrophic consequences. Given that reality, it is profoundly irresponsible for editors or reporters to pass along the story with only some counterposing quotes and without doing enough digging to satisfy themselves as to the bottom-line gravity of the situation. Their

assessment needn't be the same as that of environmentalists. But simply to treat the story like any other—without
taking the time to reach an informed judgment about its
potential gravity—is a fundamental violation of the trust of
readers and viewers who assume a modicum of informed
interpretation from their news providers.

In their paper "Balance as Bias: Global Warming and the
U.S. Prestige Press," Maxwell T. Boykoff and Jules M.
Boykoff make a strong case that the formulaic use of journalist balance has put the United States years behind the rest of
the world in beginning to act on the climate crisis.

"The continuous juggling act journalists engage in often
mitigates against meaningful, accurate and urgent coverage
of the issue of global warming," they wrote. "Since the general public garners most of its knowledge about science
from the mass media . . . the disjuncture between scientific
discourse and popular discourse [is responsible for the fact
that] significant and concerted international action has not
yet been taken to curb practices that contribute to global
warming."

On another level, slightly removed, coverage of the climate crisis has been one of many casualties of the takeover of
the news industry by a small number of massive media conglomerates. Traditionally, most newspapers were owned by
families or companies that felt a profound obligation to the
mission of news. Owners of news outlets were traditionally
content with profits of about 10 percent—as long as they
were able to fulfill what they saw as their mission of informing the public.

Unfortunately, with the acquisition of most news outlets by a small group of conglomerates, the direction of the business has been determined by the profit-driven demands of Wall Street. One result is that marketing strategy is replacing news judgment. Another result is that most newspapers have been cutting staff and failing to provide reporters with the time they need for thorough reporting of complex stories. At the same time, they have sacrificed real news coverage to increase readership and advertising through more celebrity coverage, more self-help articles, and more trivial medical news.

The result is that the complex, multifaceted, and frequently depressing story of climate change has gotten very short shrift in the news media.

There are enough aspects to the issues that surround this story—science, extreme weather, technology developments, oil industry movements, terrorism and security, diplomatic tensions, economic ramifications—that it should be in the paper three times a week. Rather than ghettoized as a sub-beat of environmental reporters, the climate issue should be integrated into much broader areas of coverage. Because it is not, the U.S. public is far less aware than most of the rest of the world of the economic and political implications of climate change.

Over and above the campaign of manufactured denial by the fossil fuel public relations specialists, there is a natural human tendency toward denial of this issue. When one is confronted by a truly overwhelming problem—and one does not see an apparent solution—the most natural human reaction is to not want to know about it. And that applies to editors just as much as readers.

For that reason, it is critical for the public to understand that there do exist solutions that would achieve the 70 percent cuts required by nature, even as they would create huge numbers of jobs and economic growth—especially in developing countries.

If a person sees that there is an intellectually honest solution, then, and only then, will he or she let the bad news in. Absent that realization, denial is the inevitable response.

The U.S. press today is in "stage-two" denial of the climate crisis. Editors acknowledge its existence even as they minimize its scope and urgency. This is evident from the pattern of coverage that provides occasional feature stories about the decimation of the forests in Alaska—but which continues to ignore the central diplomatic, political, and economic conflicts around the issue.

By underreporting this story, the press is failing to move the conversation toward solutions and, in the process, ignoring the positive potential embedded in the climate crisis. There are solutions—some of which could, if implemented, also hold the key to solving some of the most intractable problems facing humanity today.

Some observers argue that the European press has covered the climate issue more thoroughly than the U.S. press simply because European politicians raise the issue more frequently than politicians in the United States. According to that argument, the European press is no more proactive in its coverage than the U.S. press. Both are simply reflecting the agendas of their political leaders. Given the absence of any comprehensive studies on the subject, it is impossible to determine whether the disparity in coverage is simply due to the press's tendency to follow the lead of a country's newsmakers—or whether papers like the *Guardian* and the *Independent* (in Britain) and *Frankfurter Allgemeine Zeitung* and *Sueddeutsche* (in Germany) are initiating much of their coverage of the climate issue.

But in assessing the responsibility of the press, the argument seems somewhat academic. If a political leader raises an issue, the press follows it. Conversely, if the press raises a significant issue, it is almost impossible for politicians to ignore it. Witness the political responses to the press coverage in late 2003 of mad cow disease—a subject that received scant coverage inside the United States prior to its emergence in an infected cow in Washington state. That coverage put the issue of food safety into the political agenda in the United States.

The power of the press in the United States, however diluted by commercial pressures, is still formidable. When the press covers an issue thoroughly and consistently, the pubic responds. Policies are changed. Laws get passed. Witness, for example, the press's coverage of tobacco science, which has profoundly changed the nation's smoking habits. Press stories a generation ago, which highlighted instances of discrimination against African Americans and other ethnic groups, contributed to major changes in the country's civil rights laws. In the 1970s, stories about the degradation of the Great Lakes and the increasing dangers of chemical contamination prompted then president Richard Nixon to create the Environmental Protection Agency. Press revelations about massacres in Vietnamese villages and official underreporting of U.S. casualties contributed to a tidal wave of protest that ultimately led to America's withdrawal from Vietnam. And the follow-up coverage of an apparently unremarkable break-in in a Washington, D.C., office led to the impeachment and resignation of a president of the United States. One would think that the climate crisis, in all its unprecedented peril and promise, merits at least the same degree of media attention.

Finally, the climate issue is riven with conflicts at every level—and conflict is, if nothing else, the lifeblood of journalism. This issue, moreover, presents a tremendous opportunity for professional gratification. To sidestep this story is to deprive oneself of an extraordinary professional challenge. This is an immense drama. Its outcome is very much in doubt. This is by far the most important and exciting story any reporter could ever want to work on.

The conflicts are there. They are just waiting to be written.

Snapshots of the Warming No. 4

Separate and apart from the warming of surface waters, our fossil fuel burning is affecting the world's oceans in at least three significant ways. It is changing the pattern of El Niños, which play havoc with weather all over the world. The fallout from our carbon emissions, moreover, is increasing the acidity of the earth's oceans. But the most potentially catastrophic oceanic changes may result from warming-driven disruptions in the normal flow of deep-water currents that determine climatic conditions in much of the world.

Start with the increase in intensity and frequency of El Niños, which alter the world's weather patterns in very dramatic and destructive ways. El Niños are basically the oceans' mechanism for expelling stored heat. Heat accumulates in the oceans, is recycled through a complex system of currents toward the poles, and then surfaces periodically as a warm pool in the western Pacific. The heat is then released into the atmosphere—and the El Niño is normally replaced by La Niña, which is marked by the emergence of a cooler pool of Pacific surface waters.

As Kevin Trenberth explained in his 1999 paper, "The Extreme Weather Events of 1997 and 1998," El Niño events are "most simply an alternating sequence of storing up and then releasing thermal energy—like repeatedly filling a bucket, then pouring most of it out. The spacing of El Niño events would then be determined by the time spent in recharging the system—that is, in accumulating a sufficient volume of warm water in the tropics—plus the time for the ensuing El Niño to run its course. An endless cycle of charging and then discharging would also explain why El Niños are preceded and followed by La Niñas."

But the historic pattern of El Niños—which surfaced on average about every seven years—changed dramatically in the mid-1970s, with El Niños becoming more frequent and more intense.

According to Trenberth: "The Warm Pool in the tropical western Pacific is expanding; and/or the recharge phase of El Niño has speeded up; and/or the heat loss phase is less efficient. Any of these could follow from [atmospheric] warming and result in more frequent El Niño events. With global greenhouse warming we should expect higher temperatures in the upper layers of the ocean, and a steeper drop in temperature beneath the surface, which would increase the magnitude of swings between La Niña and El Niño."

"It may prove significant that the emergence of clear signs of global surface warming, beginning in the late 1970s, is coincident in time with the more frequent appearance of El Niños," wrote Trenberth in his groundbreaking paper in Consequences *in 1999.*

Two years later, an international team of scientists, using fossilized coral reefs, reconstructed the history of El Niños over the past 130,000 years.

"This situation makes it possible to sample corals which grew during periods when climatic boundary conditions were substantially different from those today," the researchers wrote in their paper.

"The samples indicated that El Niño was never more intense than the events of the last hundred years," concluded David Lea, a professor of geological sciences at the University of California–Santa Barbara, who helped author the study. In fact, the El Niño of 1997–1998 was the most severe on record.

In studying its impacts, Trenberth noted some of the weather extremes it triggered: extensive flooding in Korea in August 1998, and in eastern India and Bangladesh the following months. The flooding also released China's Yangtze River from its banks, killing more than 3,000 people, leaving 230 million people homeless, and generating over $30 billion in damages. *"The costliest disaster of them all in terms of human life,"* Trenberth wrote, *"struck the Caribbean in late October. Hurricane Mitch caused the deaths of more than 11,000 people in Honduras, Nicaragua, Guatemala, and El Salvador, primarily through the extensive flooding that followed prolonged and heavy rains."*

Even as they are heating up, the world's oceans are also becoming increasingly acidified—due to their absorption of carbon dioxide. Two researchers at the Lawrence Livermore National Laboratory reported in September 2003 that the pH levels of the world's oceans were changing with increasing rapidity. *"If we continue down the path we are going, we will produce changes greater than any experienced in the past 300 million years,"* said one of the researchers, Kenneth Caldeira.

Their findings indicated that after the last ice age, the pH of the oceans was 8.3. It had changed to 8.2 by the beginning of the industrial era. That change took about 10,000 years. But following that, in little more than a century, the pH dropped to 8.1.

"We are changing the chemistry of the ocean and we don't know what it's going to do," Caldeira told the New Scientist. *But it is clear that a change in ocean chemistry threatens the health of every kind of marine life—from plankton to coral reefs to fish.*

Of all the ocean-related impacts of climate change, nothing presents a more frightening scenario than a warming-driven change in deep-ocean currents. This change carries with it the potential for a rapid climate change event that could, paradoxically, plunge much of the overheated Northern Hemisphere into a deep freeze.

"Projected climate changes during the 21st century have the potential to lead to future large-scale and possibly irreversible changes in Earth systems resulting in impacts at continental and global scales . . . Examples include significant slowing of the ocean circulation that transports warm water to the North Atlantic," among other scenarios.

Were the ocean circulation to slow, the result could be a rapid and catastrophic deep freeze in northern Europe and northern North America—where the climate is tempered by the warmth conveyed by ocean currents. This type of cataclysmic climate change event took place about 11,000 years ago, changing the climate in Britain to that of the climate in Greenland. According to scientists who have read the ice-core records from that period of time, the entire change took place in about four years.

(The natural warming that took place near the end of the last ice age is thought to have occurred because as the ice cover thickened, ocean levels dropped. As a result, they exposed frozen deposits of methane hydrates that subsequently thawed into a gaseous form, rose into the atmosphere, and began to trap in heat, according to Barrett Rock, director of the Complex Systems Research Center at the University of New Hampshire. The resulting warmer atmosphere began to melt the ice cover, releasing a flow of fresh water into the North Atlantic.)

"The very recent freshening signal in the North Atlantic is arguably the biggest and most dramatic change in ocean property that has ever been measured in the global ocean. Already, surface waters in the Greenland Sea are sinking at a rate 20 percent slower than in the 1970s," wrote Robert Gagosian, head of the Woods Hole Oceanographic Institution, in 2003. "At what percent will the Ocean Conveyor stop? 25 percent? 40 percent? 60 percent? This is not like a dimmer switch, but more like a light switch. It probably goes from 'on' to 'off.' We can't yet determine the precise source or sources of this additional fresh water. Global warming may be melting glaciers or Arctic sea ice. In recent decades, the volume of Arctic sea ice has decreased by 40 percent. And if North Atlantic sinking slows down, less salty Gulf Stream waters flow northward—which exacerbates the situation. In February 2002, at a worldwide meeting of oceanographers, new data on North Atlantic freshening prompted many scientists to say that salinity levels in the North Atlantic are approaching a density very close to the critical point at which the waters will stop sinking."

At the end of 2003, researchers reported in the journal Nature *that the salinity of the North Atlantic was decreasing. Using salinity measurements from the bottom of Greenland to the tip of South America, the researchers, led by Ruth Curry at Woods Hole Oceanographic Institution, discovered that warming-driven increased rates of evaporation had increased the saltiness of tropical waters by 10 percent over the previous fifteen years. At the same time, they found a corresponding freshening of water in the northern and far southern Atlantic Ocean. In addition to the change in salinity, the increase in evaporation is putting more water vapor into the atmosphere, which, in turn, traps in more warming. Concluded Curry: "The system appears to be revving up."*

Added Ransom Myers, professor of ocean studies at Dalhousie University in Halifax: "We're changing the Earth in ways that are just inconceivably large."

As Terry Joyce, a colleague of both Curry and Gagosian at Woods Hole, noted: "I'm in the dark as to how close to an edge or transition to a new ocean and climate regime we might be. But I know which way we are walking. We are walking toward the cliff."

A subsequent study of ocean circulation changes in early 2004, spearheaded by Nobel Prize–winning scientist Paul Crutzen, sounded an even more urgent note. As the British newspaper the Independent *headlined the findings: "Global Warming Will Plunge Britain into New Ice Age 'Within Decades.'"*

5

Three Fronts of the Climate War

We are all adrift in the same boat. And there is no way half the
boat is going to sink.
℘ Raul Estrada Oyuela, Argentine
climate negotiator, Kyoto, Japan
December 1997

Although the battle over the climate issue is most vividly il-
lustrated by the relentless resistance of big coal and big oil
within the United States, it has rippled throughout the political, diplomatic, and business arenas—pitting nations and
industries against each other and even setting the federal
government against many states.

Within a month of taking office, President George W.
Bush opened a gaping rupture between the United States
and Europe on an issue of paramount importance to the
Europeans—global climate change.

That split over the climate crisis would be reflected in
growing divisions between the United States and the rest of

the world, between Washington and many U.S. state and city governments, and within the business world as well, exposing deep differences within the auto, oil, and insurance industries.

The Bush administration's diplomatic posture mirrored one of its central ideological goals: the drastic reduction of the power and influence of government domestically—and the concurrent reduction of the influence and reach of international governance institutions.

Bush aroused the suspicions of many U.S. allies when, a month after his inauguration, he reversed his campaign promise to cap emissions of carbon dioxide from power plants. Under pressure from lobbyists in the coal industry, as well as from conservative members of the Republican Party, Bush announced on March 13, 2001, that he would no longer seek to regulate such power plant emissions.

The statement dismayed many Democrats—and a number of Republicans, including then treasury secretary Paul O'Neill, a strong proponent of aggressive climate policies. But the strongest negative response came from across the ocean.

Nine days after his announcement, Bush received a stern letter from the fifteen-nation European Union condemning his action. The letter, signed by European Commission president Romano Prodi and Swedish prime minister Goeran Persson, challenged Bush to find the "political courage" to tackle the climate crisis.

The letter made it clear that to the EU, an agreement "leading to real reductions in greenhouse gas emissions is of the utmost importance . . . The global and long-term importance of climate change and the need for a joint effort by all industrialized countries in this field makes it an integral part of relations between the USA and the EU."

The president's response to the EU was unequivocally dismissive. Six days after receiving the letter from the EU, Bush withdrew the United States from the Kyoto Protocol because, in the words of the president's press secretary, "It is not in the United States' economic best interest."

Margot Wallström, the European Union environmental commissioner, called Bush's decision "very worrying." Kazuo Asakai, a top official in the Japanese embassy in Washington, told the *Washington Post*, "Japan will be dismayed and deeply disappointed. [The Kyoto treaty] is very serious and important."

The European diplomats were particularly stunned by the fact that the administration had failed to inform them of its plan before announcing it to the media.

"Sometimes people think this is only about the environment, but it's also about international relations and economic cooperation," EU spokeswoman Annika Ostergren told Reuters News Service. "The EU is willing to discuss details and problems, but not to scrap the whole protocol."

Swedish prime minister Goeran Persson was sharply critical of Bush, telling reporters that Bush's position was a heavy blow to the international effort to curb global warming. "It will have a tremendous impact . . . because it would have sent an extremely strong signal if the U.S. had stuck with the Kyoto protocol," Persson said.

Ironically, just weeks earlier, Bush's EPA administrator, Christine Todd Whitman, had warned him in a memo that he must demonstrate his commitment to cutting greenhouse gases or risk undermining the standing of the United States among its allies. "Mr. President, this is a credibility issue for the U.S. in the international community. It is also an issue that is resonating here at home," she wrote in a March 6, 2001, memo. "We need to appear engaged."

Her fears were well founded. Bush's withdrawal from the Kyoto process created the impression that Bush "[is] hawkish, unilateralist and may appear in some areas isolationist," according to Helmut Sonnenfeldt, a European affairs specialist at the Brookings Institution.

That perception moved from the diplomatic corridors to the streets a few months later when tens of thousands of demonstrators mounted a massive protest in Genoa against the Bush withdrawal from the Kyoto process.

The demonstrators' anger at the United States was noted by several foreign leaders, including a skeptical Jean Chrétien, prime minister of Canada. Noting Bush's promise to provide an alternative plan to the Kyoto mechanism, Chrétien said he would listen to an American proposal but "was not waiting for it." Regardless of the substance of any proposal by the Bush administration, Chrétien vowed that Canada would proceed to ratify the protocol.

Added French president Jacques Chirac: "The elected leaders of our countries have to consider the problems that have brought tens of thousands of our compatriots, mainly from European countries, to demonstrate their concern, to demonstrate their wish to change."

Despite U.S. intransigence, however, the Europeans continued to pursue a policy to slow global warming. In 2002, the EU voted to ratify the Kyoto Protocol, and European commissioner Margot Wallström warned that the pressure was now on the United States, the world's biggest single emitter of greenhouse gases, to step up to the challenge. "The European Union urges the United States to reconsider its position," she said.

The ratification by the European Union followed a declaration, eighteen months earlier, that the EU would get most of its reductions from new renewable energy installations rather than from the dubious and less reliable mechanism of international emissions trading. In November 2000, French environment minister Dominique Voynet declared that the core position of the EU was to ensure that countries made most of their emissions cuts through domestic action rather than through trading emissions credits. She demanded a tight limit on the use of so-called carbon sinks—forests, grasslands, and other forms of vegetation that absorb carbon—for countries to meet emissions reduction targets.

That relatively uncompromising position set the tone for a major conference in Johannesburg two years later, the World Summit on Sustainable Development (WSSD). In preparing for the summit, leaders of countries in Europe and elsewhere declared that a global switch to clean energy needs to be at the heart of any major reversal of the growing—

and perhaps irreversible—environmental degradation of the planet.

Their agenda, as the WSSD approached, centered on plans both to boost renewables and to revive what was already quickly becoming a moribund Kyoto process.

But the United States made clear a month before the summit what its position would be. As Senator James Jeffords (I–VT) made clear, the White House would try to "keep global climate change off of the agenda" at the Johannesburg meeting. Jeffords, a Vermont Republican turned Independent who chaired the Senate's environmental panel, bemoaned the fact that the administration would send a small, lower-level delegation to Johannesburg in an effort to "narrow the scope of the discussions."

That prediction was realized in September, when Bush refused to attend the 191-nation summit—despite the presence of an array of world leaders, and despite the fact that a predecessor summit in 1992 in Rio had been attended by his father, President George H.W. Bush.

Not only did George W. Bush refrain from attending, he also saw to it that the plans of other nations to launch a major clean-energy initiative were thwarted by the United States and some of its oil-producing allies. In the run-up to Johannesburg, a number of countries, led by the EU, put the issue of renewable energy near the top of the world's sustainability agenda. But near the end of the meeting, the United States, acting through the Saudi Arabian and Venezuelan delegations, pushed through a resolution that would maintain coal and oil as the world's primary fuels.

"We have lost an opportunity to move forward substantially on renewable technologies internationally," said Michael Marvin, executive director of the U.S. Business Council for Sustainable Energy. "If we're truly looking to change patterns of investment, this doesn't do it."

After a week of deadlocks and multiple drafts, the delegates finally agreed on the role of energy in future development. But the final resolution endorsed such unsustainable technologies as clean coal and large-scale hydropower. Most tellingly, it required no deadlines or timetables for action.

The consequences of the Bush administration's behavior—on this and other fronts—may well have far-reaching diplomatic, as well as climatic, consequences.

"The U.S. decision on Kyoto could become a turning point in trans-Atlantic relations," wrote Jessica T. Mathews. In an article in *Foreign Policy*, Mathews, who is president of the Carnegie Endowment for International Peace, noted that the Bush administration turned its back not only on Kyoto but also on four other major international compacts: the International Criminal Court, the ban on antipersonnel land mines, the biodiversity treaty, and the Biological Weapons Control Treaty.

In refusing to support these compacts, the United States was ignoring the accelerating integration of Europe, she asserted. "Economically, the EU is no longer a junior partner. It has a larger population than the United States, a larger percentage of world trade, and approximately equal gross domestic product," she wrote at the end of 2001.

Climate change is the leading edge of an increasing number of problems that are truly global in scope. And these issues create a need for new rules that "nibble away at the edges of national sovereignty"—a trend that runs directly counter to the neo-nationalism of the Bush administration.

This "à la carte multilateralism"—in which the United States decides which issues it is willing to cooperate on—"is not an approach that goes down easily" in the rest of the world, Mathews wrote.

Her conclusion: The current U.S. posture, which is epitomized by the U.S. position on climate change, could well result in the loss of America's position as global political leader.

"America's interests, not to mention its legitimacy and capability as a world leader, are better served by [participating] in shaping rules and procedures rather than in sulking outside the tent. Though Europe cannot challenge U.S. political or military supremacy, the world's single superpower must acknowledge that its power no longer translates . . . into a community of Western democracies and Third World dependents ready to fall into line behind U.S. leadership."

Mathews's article, written shortly after the terror attacks in September 2001, apparently fell on deaf ears.

With the Bush administration's success in assembling a coalition against terrorism—and its subsequent "coalition of the willing" to support the war in Iraq—the administration has, in fact, been quite successful in "cherry picking" issues on which it wants international agreement.

In those areas in which it opposes such agreement—such as global climate change—it has fashioned a very different policy.

In the case of climate change, the Bush administration has set about to destroy the Kyoto process through a series of bilateral agreements with individual countries. In mid-2002, one month after withdrawing from the Kyoto process, Australia signed a bilateral agreement with the United States to cut emissions—an agreement that most observers believe will do nothing to avert climate chaos. As Dr. Frances Maguire of Greenpeace Australia pointed out: "The agreement lacks any calls for direct action. It does not require any reduction in greenhouse gas emissions."

Sunita Narain, director of the Centre for Science and Environment in New Delhi, summed up the administration's strategy in late 2003. "The U.S. walked out of the Kyoto Protocol. It walked out of a multilateral agreement to limit luxury emissions, so that the poor would get ecological space, and the earth's climate system would recover. But the U.S. did not merely reject the protocol; it said it would work overtime to kill it off. The U.S. says it will prove its strategies for 'voluntary measures'—to switch to cleaner technologies—and 'bilateral agreements' will be more effective than a multilateral rule-bound agreement. Forget the rules now. What the U.S. is promising is that instead of the 5.2 per cent cut in emissions at 1990 levels, as the protocol requires, it will increase its emissions by over 30 per cent in the agreement period.

"The U.S. is certainly aggressive about its ideology. Its diplomatic cohorts go around the world wrapping up deals—on hydrogen technologies, carbon sequestration plans, or clean coal. Earlier this year, a U.S. climate delegation neatly stitched up the so-called hero of the developing world, India. It wowed our politicians and business leaders with a grandiose and futuristic hydrogen energy plan, in which hydrogen will be generated using fossil fuels (oil from Iraq?) as the source."

In fact, U.S. delegations also met with representatives in Thailand, Malaysia, and Vietnam to pursue individual treaties and further undermine a coordinated international approach based on multilateral agreements.

A subsequent bilateral agreement with Italian president Silvio Berlusconi, Narain wrote, involves a joint program "on so-called climate change mitigation strategies. Just two months before the next climate conference—which the Italians are hosting—the two have thumbed their noses at the multilateral world with public advertisements of this marriage."

As the nations of the world gathered in Milan at the beginning of December 2003 for the ninth round of Kyoto talks, the Bush administration made its feelings very clear.

Kyoto is "an unrealistic and ever-tightening regulatory straitjacket, curtailing energy consumption," Paula Dobriansky, U.S. undersecretary of state for global affairs, wrote in the *Financial Times* newspaper. Added Conrad Lautenbacher, an undersecretary of commerce in the Bush administration, in a press conference in Milan: "Where we are today in climate change science is problematic. We need more fundamental understanding."

The Bush administration's desire to drive a stake through the heart of the Kyoto process may have been realized on December 2, 2003, when Russian president Vladimir Putin, after extensive conversations with the Bush administration, indicated strongly that Russia would most likely not ratify the protocol.

That announcement, coupled with the U.S. withdrawal, may have effectively killed the protocol—since its taking effect required its ratification by fifty-five countries that are responsible for 55 percent of the world's carbon emissions.

The new Bush posture was a radical departure from the Clinton administration, which, even in the absence of Senate ratification, had signed the Kyoto Protocol in 1999. It reached fruition in early 2003 when President Bush, in concert with some of the nation's leading emitters of carbon dioxide, announced a plan for companies to voluntarily reduce the "greenhouse gas intensity" of their activities by 18 percent in the next ten years.

That plan was drafted with the help and approval, among others, of the Edison Electric Institute, the American Petroleum Institute, and the Southern Company, as well as the National Mining Association, the Portland Cement Association, the American Iron and Steel Institute, the American Chemistry Council, the Aluminum Association, the Association of American Railroads, and the American Forest and Paper Association.

But the sleight-of-hand underlying the plan was immediately obvious to observers. (The "intensity" of carbon emissions—which is very different from the quantity of carbon emissions—reflects the amount of CO_2 emitted per unit of economic output, not the total output of CO_2). In short order, the Washington-based World Resources Institute pointed out that given the projected increase in the nation's

economic output, the Bush plan would still allow greenhouse gas emissions to increase by about 14 percent during
the period of the plan—only marginally less than would
have been expected in the absence of a mandatory, let alone
voluntary, intensity-reduction arrangement.

Europe's anger at the Bush administration's disingenuous
and deceptive approach erupted again several months later
when Britain's top science adviser launched a public attack on
U.S. priorities. In an article in *Science,* Sir David King point-
edly asserted that "climate change is the most severe problem
that we are facing today, more serious even than the threat of
terrorism." In a withering assault, he blasted Bush's "free-
market" approach to emissions reductions, declaring: "The
Bush administration's strategy relies largely on market-based
incentives and voluntary action . . . But the market cannot
decide that mitigation is necessary, nor can it establish the basic international framework in which all actors can take their
place." Concluded King: "As the world's only remaining
superpower, the United States is accustomed to leading internationally co-ordinated action. But the U.S. government is
failing to take up the challenge of global warming."

⚘

Even as the Bush approach alienated other countries, it also
set the federal government against many of the country's
states and cities.

In the absence of federal leadership, a number of locali-
ties decided to take the first step in addressing the climate

crisis, confident that the national government would soon follow suit. But as the Bush administration's antagonism toward this issue became increasingly apparent, a number of states moved the battle from the grass roots into the courts.

Picking up on a voluntary, local approach begun in 1988, more than 110 cities have now joined with the Cities for Climate Protection Campaign of the International Council for Local Environmental Initiatives. More than thirty states have either developed—or are already implementing—programs to reduce their carbon emissions.

In late 2002, the Pew Center for Climate Change reported that more than half the states had enacted mandatory or voluntary plans to reduce emissions. According to a report in the *Washington Post* that was based on the Pew study, fifteen states, including Texas, have enacted legislation requiring utilities to increase their use of renewable energy sources such as wind power or biomass in generating a portion of their overall electricity.

The *Post* cited a number of examples, including:

- New regulations in California to reduce car and truck emissions by 2006
- A Texas requirement that 3 to 4 percent of its electricity will come from renewable energy sources, especially wind power, by the end of the decade
- A formal target in Massachusetts to reduce power-plant emissions of carbon dioxide
- A New Hampshire law to limit emissions for the state's three aging coal-fired electric generating plants

- Nebraska legislation promoting the planting of trees to increase the state's absorption of carbon dioxide
- A New Jersey commitment to lower greenhouse gas reductions by 3.5 percent below 1990 levels, by 2005, including a surcharge on consumers' utility bills of $358 million for energy efficiency and renewable energy

As their frustration with the president intensified, a number of states became more aggressive in their criticisms of the Bush administration and its Environmental Protection Agency for turning their backs on the climate crisis.

In July 2002, the attorneys general of eleven states called on the White House to provide leadership in combating climate change. Calling global warming "the most pressing environmental problem of our time," the attorneys general called on Bush to cap carbon emissions. The group, which included the chief legal officers of New York, California, Massachusetts, and Alaska, criticized the president for not proposing "a credible plan that is consistent with the dire findings and conclusions being reported."

"By acting now to reduce greenhouse gas emissions, the Bush administration can provide regulatory certainty to the business community, can spur private sector investment in renewable energy and energy efficiency, and can lay the groundwork to avoid the potentially disastrous environmental, public health, and economic impacts of global warning," noted New York attorney general Elliot Spitzer.

Several months later, at the U.S. Conference of Mayors, the chief executives of nearly 250 cities called on Bush to act now on climate change, noting: "[T]he scientific community has reached a consensus that human activities are impacting the Earth's climate."

They added, "Mayors are uniquely situated to lead national climate protection efforts by taking action in a broad range of areas." In their resolution, they noted that "many mayors are already pursuing programs and policies to reduce greenhouse gas emissions in their cities and communities, including more than 125 local governments that have committed to assessing emissions, setting a specific reduction target for greenhouse gas emissions and monitoring progress."

When the Bush administration refused to heed the calls of state and city officials, the attorneys general of seven states escalated their efforts in early 2003, filing a lawsuit against the federal government for refusing to regulate carbon dioxide through the Environmental Protection Agency.

They dropped that suit in favor of a broader one eight months later when eleven states, along with the District of Columbia and American Samoa, sued the federal government for its failure to deal with climate change. That suit, which was still pending as of this writing, claimed the EPA is required by the U.S. Clean Air Act to regulate greenhouse gas emissions.

"Because the United States is already dealing with the harmful effects of global warming, the American people want less talk and more action now," Rhode Island attorney general Patrick Lynch said in a statement.

The attorneys general dismissed a claim by the EPA that the agency lacked authority from Congress to regulate greenhouse gases, citing the EPA's denial of a petition to impose controls on vehicles' greenhouse gas emissions.

Massachusetts attorney general Thomas Reilly said the carbon emissions are causing real environmental and health problems. "You're seeing the erosion of our beaches. You're seeing saltwater contaminate our drinking water. You see damage to our infrastructure, to our roads and our causeways and our bridges," he said.

On the West Coast, the governors of Washington, Oregon, and California announced in 2003 that they were pooling their resources to buy high-efficiency vehicles, develop renewable sources of electricity, and institute a verifiable system of measuring and reporting on greenhouse gas emissions.

Commenting on the partnership, K. C. Golden, of the Washington-based group Climate Solutions, noted: "We can't afford to wait while the federal government fiddles. We have too much to lose as the climate becomes unstable, and too much to gain by taking a leadership role in developing climate solutions. The rest of the world's advanced economies have already begun to retool for a successful, prosperous transition to clean energy sources and efficient energy systems. With this announcement, the Governors are clearly signaling that the federal government won't stop America's most forward-looking states from taking action."

❧

The divisions between the United States and Europe and between Washington and the rest of the country are reflected in divisions within a number of industries—most notably the oil, auto, and insurance industries.

The first major oil company to acknowledge the urgency of climate change was British Petroleum, whose CEO, John Browne, announced in 1997 that the company was preparing itself to begin the transition to renewable energy in order to reduce the world's output of greenhouse gases.

The company subsequently launched a series of advertisements in which it declared that its initials no longer stood for British Petroleum but for "Beyond Petroleum." Observers credited the company with running those ads not on PBS—but during the baseball championship playoff series, aiming them clearly at a mainstream male American audience.

Beyond reshaping its public image, BP also spawned a new subsidiary—BP Solar—which has grown into the world's largest seller of solar systems.

Another oil giant, Royal Dutch/Shell, announced in 2001 that the industrial world could be getting half its energy from renewable energy and natural gas by 2020. (Natural gas emits only half as much carbon as coal per unit of energy and about two-thirds as much as oil.)

Shell followed that announcement by doubling its existing $500 million investment in renewables to $1 billion. The company said the continued investment would focus mainly on solar and wind power.

By contrast, the world's biggest oil company, Exxon-Mobil, with its extraordinarily close ties to the Bush White

House, has continued to counteract the initiatives of BP, Shell, and others.

In late 2002, under pressure from shareholders, Exxon-Mobil made a token investment in research into gasoline-based hydrogen fuels. At the time, ExxonMobil declared that useful renewable energy was at least twenty years away. ExxonMobil's timetable for renewables was vividly contradicted the following day, when Toyota announced it was putting a fleet of hydrogen-powered fuel-cell cars on the streets of Tokyo.

Nevertheless, ExxonMobil's proposal to make "dirty" hydrogen (from oil and coal, rather than from nonpolluting water) was adopted as presidential policy when President Bush included it as part of his State of the Union address in 2003.

Later that year, ExxonMobil announced that, its hydrogen initiative notwithstanding, carbon emissions would grow worldwide by 50 percent by the year 2020. The company made it clear it had no intention to move away from carbon-based fuels.

Despite the fact that humanity is now putting about 7 billion tons of heat-trapping carbon into the atmosphere every year, several top officials of ExxonMobil emphasized in a private conversation that the company has no intention of reducing its carbon output. As ExxonMobil's president, Lee Raymond, said in a recent interview: "The mainstream of some so-called environmentalists or politically correct Europeans isn't the mainstream of all scientists or the White House. The world has been a lot warmer than it is now and it didn't have anything to do with carbon dioxide."

The split within the carbon industry was underscored in the fall of 2003 when a top official of one of the country's largest utility companies blasted the Bush administration's position. "I'm a lifelong Republican and an admirer of much of what President Bush has accomplished," said Robert Luft, chairman of the board of the Entergy Corp. "Still, I cannot express my frustration with his performance in this area. . . . Make no mistake. If today's leaders of government and business don't start understanding the need to take emission reductions seriously, we will leave a grim, grim legacy for our children and grandchildren."

The responses of the insurance industry have been equally schizophrenic. The big European insurers have been politically proactive. In the early rounds of the climate talks, they aligned themselves with a coalition calling for the largest initial cuts (20 percent below 1990 levels)—the Alliance of Small Island States, from Jamaica to the Philippines, countries whose stability is threatened by rising sea levels and increasingly intense storm surges. The European insurers have also spent large amounts on public education, newspaper advertising, and political capital on the climate threat.

By contrast, most U.S. insurers have been economically defensive and politically invisible. Insurers in this country have withdrawn coverage further and further inland from coastlines. They are refusing to insure known storm corridors and selling the risk off to the public. They are keeping silent politically.

The concern of the European insurers is reflected in their estimates of coming economic losses. The United Nations Environmental Programme (UNEP) has projected that climate damages will amount to $150 billion a year within this decade. The world's largest insurer—Munich Reinsurance— has said that within several decades, those losses will amount to $300 billion a year. And two years ago, Britain's biggest insurer projected that, unchecked, climate change could bankrupt the global economy by 2065—from property damage due to sea level rise and increasingly severe storms and floods; destruction of energy, health, and communications infrastructures; crop failures; losses in the travel and tourism industries; and public health costs.

In the spring of 2003, Swiss Reinsurance, one of the world's largest reinsurance companies, took a major step beyond corporate rhetoric into corporate operating policy. The company began to ask its directors and officers insurance clients what they were doing to prepare for government restrictions on their emissions of greenhouse gases. According to the *Wall Street Journal,* "Swiss Re is considering denying coverage, starting with directors-and-officers liability policies, to companies it decides aren't doing enough to reduce their output of [greenhouse] gases."

Directors-and-officers liability coverage protects named officers and directors from personal liability arising from corporate mismanagement—the kind of suits that have swirled in the wake of recent scandals involving companies like Enron, WorldCom, and Tyco. In this case, the liability would arise not from corporate fraud but from failure to deal with the coming consequences of climate change.

"Emissions reductions are going to be required. It's pretty clear," said Christopher Walker, a managing director of Swiss Re. "So companies that are not looking to develop a strategy for that are potentially exposing themselves and their shareholders."

The world's largest reinsurer, Munich Reinsurance, is proceeding down the same road as Swiss Re. One executive of Munich Re noted that that company, too, may begin to deny coverage to firms that are not preparing to reduce their greenhouse gas emissions.

One of Swiss Re's clients, Baxter International, Inc., a major producer of health-care equipment, decided six years ago to begin publicly reporting their greenhouse gas emissions. "Global warming will be increasingly detrimental to the well-being of millions," the company says on its Web site. "All large companies must step forward and . . . address this critically important issue."

A Baxter vice president, responding to the new Swiss Re coverage policy, said: "When the insurance companies are debating things, they're debating them because they're beginning to see there may be practical consequences. And when that happens, you've got to pay attention."

Following up on its demands that its clients attend to their greenhouse gas footprints, Swiss Re is also promoting the use of emissions trading as a means to facilitating the reduction of carbon emissions. As a leading proponent of greenhouse gas cuts in the insurance industry, Swiss Re has become a member of the International Emissions Trading Association, the Chicago Climate Exchange, the International Centre for Carbon Sequestration (ICCS), and the Emission Market Development Group (EMDG).

Swiss Re took a further step toward mobilizing the finance world around the climate crisis in late 2003 when it sponsored, along with the United Nations Environment Programme, a conference of officials controlling some of the nation's largest pension funds.

The conference yielded a "call to action" by the treasurers and controllers of California, New York State, New York City, Connecticut, New Mexico, Oregon, Maine, and Vermont, as well as officials overseeing two major union pension funds. Noting that state treasurers and pension fund managers oversee $1 trillion in assets, the group called on regulators and business leaders to force corporations to give investors more information on the financial risks from global climate change.

Said California treasurer Philip Angelides: "In global warming, we are facing an enormous risk to the U.S. economy and to retirement funds that Wall Street has so far chosen to ignore."

In contrast to European insurers, however, few, if any, U.S.-based insurance companies are acknowledging the threat of climate change.

In June 2002, the State Farm Insurance Company did announce that losses in its home insurance business were prompting it to limit or halt the sale of new policies in twenty states, due to storms, floods, and other extreme weather events. But the company's announcement failed to acknowledge that any connection between the heating of the atmosphere and the increase in destructive weather extremes was forcing it to cut back on its areas of coverage.

Gene Lecomte was a former president of the Institute for Business and Home Safety as well as CEO of the Insurance Institute for Property Loss Reduction, a consortium of large U.S.-based insurers. He outlined the differences between European and U.S. insurers at a January 2003 conference at Tufts University. "In Europe, insurers are involved in formulating government policy. They are acutely aware of the potential losses from climate impacts."

Among U.S. insurers, by contrast, there is "distrust of government, a great deal of parochialism and a set of very different agendas." Insurers in the United States, Lecomte said, "continue to insist there are major uncertainties in the science. There is a great lack of knowledge about climate change among U.S. companies—and a strange kind of parochialism among stakeholders."

Rather than press for industrywide changes to address the problem, he said, U.S. insurers are simply "burying their heads" and withdrawing coverage from coastal areas, known storm corridors, and other places that are vulnerable to weather extremes. Moreover, several U.S. insurers have passed on the risks of climate-driven losses to the public in the form of so-called catastrophe bonds. The bonds, which are backed by insurers and pay double-digit dividends, seem to be attractive investment vehicles for retirees living on fixed income. But they involve considerable risk. If the insurer loses $1 billion due to a natural disaster, the bondholder loses not only the dividend but the principal as well. What many bondholders may not realize is that the United States has suffered more than forty such loss events exceeding $1 billion since 1980.

Noting that about half of all property insurance in the United States is underwritten by about twenty companies, Lecomte called on stakeholders to push the industry to address the risks of climate change. "What is really needed is for all stakeholders to figure out how to make the industry more beneficial, to make it work the way it should. And that includes not only changing corporate cultures, but also changing regulations that discourage companies from setting aside money to cover future damages and that fail to recognize the need for more research into coming climate impacts."

Any reform of the industry, he concluded, must begin with insurers seeing to it that "property owners must understand there are no free rides."

In 1997, President Bill Clinton convened a White House Conference on Climate Change, which featured, among other personages, a dazzling array of prominent business leaders. The following January, Clinton continued to focus public attention on the threat of climate change in his State of the Union address.

Today, however, the residue of corporate enthusiasm resides primarily within the Pew Center for Climate Change, under the direction of Eileen Claussen, a major climate policymaker in the Clinton White House. While the center has enlisted some high-profile corporate players, its potential seems minimized by the fact that policy changes by individual companies are limited unless there are similar commitments by all the major players within their industries.

Unfortunately, the vast majority of the nation's business leaders, intent on securing political benefits for their companies and industries, continue, as always, to take their cues from the current center of political power.

The ultimate message seems to be that the corporate community is not the place to look for political leadership. The corporate centers of innovation, productivity, and economic growth are concerned only secondarily, if at all, with public policy—most typically, only if that policy directly affects a particular company or industry. Political change does not emanate from corporate boardrooms. As BP chief John Browne wrote in 2002: "People expect big companies to act and to use their skills and access to technology to provide better choices . . . ."

But he emphasized: "The lesson is not that individual private companies can solve the problem—they cannot."

Snapshots of the Warming No. 5

There is one group of creatures for whom global warming is a boon. Of all of the systems of nature, one of the most responsive to temperature changes is insects. Warming accelerates the breeding rates and the biting rate of insects. It accelerates the maturation of the pathogens they carry. It expands the range of insects, allowing them to live longer at higher altitudes and higher latitudes. As a result, climate change is fueling the spread of a wide array of insect-borne diseases among populations, species, and entire ecosystems all over the planet.

Those diseases are already passing from ecosystems to people—and the World Health Organization now projects that millions of people will die from climate-related diseases and other impacts in the next few decades.

In 2002, a team of researchers reported that rising temperatures are increasing both the geographical range and the virulence of diseases. The implication is a future of more widespread and devastating epidemics for humans, animals, and plants.

As the Boston Globe *reported: "Researchers have long accepted that global warming will affect a wide range of organisms, but they are only now beginning to predict what those will be. While climate change scientists have studied a handful of human diseases, [this] report was the first to study dozens of diseases in both humans and nonhumans.*

"'We are seeing lots of anecdotes and they are beginning to tell a story,' said Andrew P. Dobson, professor at Princeton University's department of ecology and evolutionary biology and one of the authors. 'It's a much more scary threat than bioterrorism.'"

The researchers reported that the climate-driven spread of diseases will "contribute to population or species declines, especially for generalist pathogens infecting multiple host species. The greatest impacts of disease may result from a relatively small number of emergent pathogens. Epidemics caused when these infect new hosts with little resistance or tolerance may lead to population declines, such as those that followed tree pathogen invasions in North America during the last century."

"The most detectable effects of directional climate warming on disease relate to geographic range expansion of pathogens such as Rift Valley fever, dengue, and Eastern oyster disease. Factors other than climate change—such as changes in land use, vegetation, pollution, or increase in drug-resistant strains—may underlie these range expansions. Nonetheless, the numerous mechanisms linking climate warming and disease spread support the hypothesis that climate warming is contributing to ongoing range expansions."

"What is most surprising is the fact that climate-sensitive outbreaks are happening with so many different types of

pathogens—viruses, bacteria, fungi, and parasites—as well as in such a wide range of hosts including corals, oysters, terrestrial plants, birds and humans," wrote lead author Drew Harvell, a Cornell University biologist.

Added Dobson: "Climate change is disrupting natural ecosystems in a way that is making life better for infectious diseases. The accumulation of evidence has us extremely worried. We share diseases with some of these species. The risk for humans is going up."

"This isn't just a question of coral bleaching for a few marine ecologists, nor just a question of malaria for a few health officials—the number of similar increases in disease incidence is astonishing," added another member of the research team, Richard Ostfeld. "We don't want to be alarmist, but we are alarmed."

The risk, of course, is not confined to humans. In Canada, an explosion in the population of tree-killing bark beetles is spreading rapidly through the forests. As of late 2002, the deadly bark beetles had spread throughout an area of British Columbia nearly three-fourths the size of Sweden—about 9 million acres. Officials attributed the spread of the insects to unusually warm winters.

The massive wildfires that devastated southern California in the summer of 2003 were also made more intense by a rapid increase in the population of bark beetles that had killed large numbers of trees, turning them into tinder for the fires that blanketed the area around Los Angeles.

But the impact of the warming-driven population boom of insects on humans is likely to be at least—if not more—severe than the impact on the world's forests.

About 160,000 people currently die each year from the impacts of warming, but the World Health Organization calculates that that figure will rise into the millions in the near future—from the spread of various infectious diseases, increased heat stress, and the warming-driven proliferation of allergens.

"There is growing evidence that changes in the global climate will have profound effects on the health and well-being of citizens in countries around the world," said Kerstin Leitner, assistant director-general of the World Health Organization.

Start with the bugs. Mosquitoes, which historically could survive no higher than 1,000 meters, are now spreading malaria, dengue, and yellow fever at elevations of 3,200 meters—to populations that have never before been infected and carry no immunity to those diseases.

Mosquitoes are spreading West Nile virus—and not only throughout expanding geographical areas (as of June 2003, West Nile had surfaced in twenty-four states within the U.S.). They have also spread the disease to more than 230 species of birds, animals, humans, and other insects.

Although West Nile virus has gotten far more attention in the American press, a more familiar disease, malaria, quadrupled worldwide between 1995 and 2000. Today, mosquito-borne malaria kills at least 1 million people and causes more than 300 million acute illnesses each year. In Africa alone, malaria is killing about 3,000 children each day. "Malaria kills an African child every 30 seconds, and remains one of the most important threats to the health of pregnant women and their newborns," according to Carol Bellamy, executive director of the United Nations Children's Fund.

According to an article in Scientific American, *"Diseases relayed by mosquitoes—such as malaria, dengue fever, yellow fever and several kinds of encephalitis—are among those eliciting the greatest concern as the world warms. Mosquitoes acquire disease-causing microorganisms when they take a blood meal from an infected animal or person. Then the pathogen reproduces inside the insects, which may deliver disease-causing doses to the next individuals they bite."*

"Mosquito-borne disorders are projected to become increasingly prevalent because their insect carriers, or 'vectors,' are very sensitive to meteorological conditions," wrote Dr. Paul R. Epstein in the cover article of Scientific American. *"Cold can limit mosquitoes to seasons and regions where temperatures stay above certain minimums. Winter freezing kills many eggs, larvae and adults outright. Anopheles mosquitoes, which transmit malaria parasites, cause sustained outbreaks of malaria only where temperatures routinely exceed 60 degrees Fahrenheit. Similarly, Aedes aegypti mosquitoes, responsible for yellow fever and dengue fever, convey virus only where temperatures rarely fall below 50 degrees F."*

The problem is that there are very few areas of the planet that are cooling—and many, many areas where the temperature is rising.

Epstein, who is assistant director of the Center for Health and the Global Environment at Harvard Medical School, explained that "mosquitoes proliferate faster and bite more as the air becomes warmer. At the same time, greater heat speeds the rate at which pathogens inside them reproduce and mature. At 68°F, the immature P. falciparum *parasite takes twenty-six days to develop fully, but at 77°F, it takes only thirteen days. The Anopheles mosquitoes that spread this malaria parasite live only several weeks; warmer temperatures raise the odds that the parasites will mature in time for the mosquitoes to transfer the infection. As whole areas heat up, then, mosquitoes could expand into formerly forbidden territories, bringing illness with them. Further, warmer nighttime and winter temperatures may enable them to cause more disease for longer periods in the areas they already inhabit."*

Nor is it only warmer temperatures that propel the spread of insect-borne diseases. Weather extremes, another consequence of climate change, also play a pivotal role. *"Intensifying floods and droughts resulting from global warming can each help trigger outbreaks by creating breeding grounds for insects whose dessicated eggs remain viable and hatch in still water,"* he wrote. As floods recede, he explained, they leave puddles that in times of drought become stagnant pools. As people in dry areas collect water in open containers, these can become incubators for new mosquitoes. The insects can flourish even more if climate change or other processes (such as habitat destruction) reduce the populations of predators that normally feed on mosquitoes.

One very troubling occurrence is that malaria had been declining in the United States before the recent rapid rise in global temperatures. As Epstein explained, *"Malaria is reappearing north and south of the tropics. The U.S. has long been home to Anopheles mosquitoes, and malaria circulated here decades ago. By the 1980s mosquito-control programs and other public health measures had restricted the disorder to California."*

"Since 1990, however, when the hottest decade on record began, outbreaks of locally transmitted malaria have occurred dur-

ing hot spells in Texas, Florida, Georgia, Michigan, New Jersey, and New York (as well as in Toronto). These episodes undoubtedly started with a traveler or stowaway mosquito carrying malaria parasites. But the parasites clearly found friendly conditions in the U.S.—enough warmth and humidity, and plenty of mosquitoes able to transport them to victims who had not traveled. Malaria has returned to the Korean peninsula, parts of southern Europe and the former Soviet Union and to the coast of South Africa along the Indian Ocean," Epstein reported.

It will also, scientists report, return to Great Britain. Last year, researchers at Britain's Durham University projected that if current temperature trends persist, the United Kingdom will begin to see recurring outbreaks of malaria in the next few decades.

Nor is the phenomenon limited to malaria.

According to Epstein, "Dengue or 'breakbone' fever (a severe flu-like viral illness that sometimes causes fatal internal bleeding) is spreading as well. Today it afflicts an estimated 50 million to 100 million in the tropics and subtropics (mainly in urban areas and their surroundings). It has broadened its range in the Americas over the past ten years and had reached down to Buenos Aires by the end of the 1990s. It has also found its way to northern Australia. Neither a vaccine nor a specific drug treatment is yet available."

Another insect that flourishes in a warmer world is the tick. In coastal New England—as well as in coastal areas in Scandinavia, researchers have documented a substantial increase in tick-borne Lyme disease. The reasons: The shorter and warmer winters in the northern temperate latitudes are no longer providing the deep, prolonged killing frosts that normally kill the ticks during the winter season.

The changes in the climate affect not only infectious diseases. They are also expected to trigger far more allergies among humans. One team of researchers found that a doubling of carbon dioxide levels—which is expected to occur after 2050—produced 61 percent more pollen than normal. This, in turn, strongly suggests more virulent allergies among current sufferers and new allergies for people who were previously unaffected.

And of course there are the direct effects of heat itself.

Two years ago, the World Meteorological Organization projected a doubling of heat-related deaths in the world's cities within twenty years. "Heat waves are expected to become a major killer," World Meteorological Organization secretary general Godwin Obasi said.

That projection turned prophetic in the summer of 2003. The final death toll of that summer's heat wave in Europe approached 35,000 fatalities, according to the Earth Policy Institute.

Part of the reason for the unusually high number of heat-related deaths—which also occurred in Chicago during a heat wave in 1996, when more than 800 people lost their lives— seems to involve more than rising temperatures. It also apparently reflects the fact that greenhouse gases trap in the heat during the nighttime, preventing the normal radiational cooling that allows heat-stressed bodies to recover from the high daytime temperatures.

6

Compromised Activists

On climate change, we need to build on Kyoto but we should recognize one stark fact: even if we could deliver on Kyoto, it will at best mean a reduction of 1 per cent of global warming. But we know . . . we need a 60 per cent reduction worldwide. In truth, Kyoto is not radical enough.

✍ BRITISH PRIME MINISTER TONY BLAIR
SEPTEMBER 2, 2002

Confronted by the steel wall of resistance of the fossil fuel lobby and their political allies, most climate activists have retreated into approaches that are dismally inadequate to the magnitude of the challenge.

Around the country, environmental advocates are working to get people to drive less, turn down their thermostats, and reduce their energy use. Unfortunately, though many environmental problems can be addressed through lifestyle changes, climate change is not one of them.

Several of the country's leading national environmental groups are promoting limits for future atmospheric carbon levels that are the best they think they can negotiate. Although those carbon levels may be politically realistic, they would likely be climatically catastrophic.

Most advocates, moreover, are relying on goals and mechanisms that were proposed about a decade ago, before the true urgency of the climate crisis became apparent. Activists—both in Washington and around the country—continue to push for the United States to meet its initial goal under the Kyoto Protocol—carbon emissions cuts of 7 percent below 1990 levels. (The protocol's aggregate goal for industrial nations is a cut of 5.2 percent below 1990 levels, by 2012.)

But those goals have been rendered obsolete by the escalating pace of climate change.

Virtually all the approaches by activists in the United States, moreover, are domestic in nature. They ignore both the world's developing countries and, equally important from the standpoint of national security, the oil-producing nations of the Middle East. Ultimately, even if the United States, Europe, Canada, Australia, and Japan were to cut emissions dramatically, those cuts would be overwhelmed by the coming pulse of carbon from India, China, Mexico, Nigeria, and all the other developing countries struggling to stay ahead of poverty.

Many alternative approaches, moreover, rely on market-based solutions because their proponents believe that in an age of market fundamentalism, no other approach can gain political traction. Unfortunately, nature's laws are not about supply and demand. Nature's laws are about limits, thresh-olds, and surprises. The progress of the Dow does not seem to influence the increasing rate of melting of the Greenland Ice Sheet; the collapse of the ecosystems of the North Sea will not be arrested by an upswing in consumer confidence.

Many groups justify the minimalist goals of making people more energy efficient as the first phase in building a political base for more aggressive action. In the past, that pattern has been successful in developing various move-ments. In the case of climate change, however, nature's timetable is very different from that of political organizers. Unfortunately, the signals from the planet tell us we do not have the luxury of waiting another generation to allow for the orderly maturation of a movement.

Finally, the environmental establishment insists on cast-ing the climate crisis as an environmental problem. But cli-mate change is no longer the exclusive franchise of the environmental movement. Any successful movement must involve horizontal alliances with groups involved in interna-tional relief and development, campaign finance reform, public health, corporate accountability, labor, human rights, and environmental justice. The real dimensions of climate change directly involve the agendas of a wide spectrum of ac-tivist organizations.

Regrettably, the environmental movement has proven it can not accomplish large-scale change by itself. Despite occa-sional spasms of cooperation, the major environmental groups have been unwilling to join together around a unified climate agenda, pool resources, and mobilize a united campaign on the climate. Even as the major funders of climate and energy-oriented groups hold summit meetings in search of a common vision, they shy away from the most obvious of imperatives: using their combined influence and outreach to focus attention—and demand action—on the climate crisis. As the major national groups insist on promoting exclusive agendas and protecting carefully defined turf (in the process, squandering both talent and donor dollars on internecine fighting), the climate movement is spinning its wheels.

The solution to the climate crisis involves a high-stakes battle with big coal, big oil, and the immense financial resources and political levers at their disposal. Soft approaches do not normally prevail in hardball competition.

Take the critical issue of climate stabilization—the level at which the world agrees to cap the buildup of carbon concentrations in the atmosphere. The major national environmental groups focusing on climate—groups like the Natural Resources Defense Council, the Union of Concerned Scientists, and the World Wildlife Federation—have agreed to accept what they see as a politically feasible target for 450 parts per million of carbon dioxide. (Our relatively hospitable climate for the past 10,000 years depended on a level of 280 parts per million.)

While the 450 goal may be politically realistic, it would likely be environmentally catastrophic. With carbon levels having risen by only 90 parts per million (from their pre-industrial level of 280 ppm to more than 370 ppm today), glaciers are now melting into puddles, sea levels are rising, violent weather is increasing, and the timing of the seasons has changed—all from a 1°F rise in the last century. Carbon concentrations of 450 ppm will most likely result in a deeply fractured and chaotic world—if not a world rendered unrecognizable by a devastating and abrupt "climate snap" that could plunge the Northern Hemisphere into a deep freeze within a decade.

(Ironically, though most environmentalists are too timid to raise alarms about so nightmarish a climate threat, the country's national security establishment is actively preparing for the kind of large-scale political destabilization it would unleash. Anticipating the possibility of a rapid "climate snap," a new Pentagon scenario envisions far more violent storms and more megadroughts in continental interiors. Pentagon planners describe masses of refugees from Mexico, South America, and the Caribbean swarming U.S. borders in search of food. In Europe, they foresee the deep freeze propelling large numbers of people from Scandinavia, Germany, and other parts of northern Europe into Spain, Italy, and points south. China's food supplies, they note, would be devastated by increasingly intense monsoons and droughts. Much of Bangladesh would become uninhabitable as rising levels of saltwater contaminate inland sources of fresh water. And all over the world, countries would be drawn into wars over dwindling amounts of arable land, shrinking supplies of potable water, and increasingly scarce parcels of climatically hospitable territory, according to Pentagon planners.)

The major, national environmental groups, moreover, are trapped in a "Beltway" mentality that measures progress in small, incremental victories. They are operating in a

Washington environment that is, at best, indifferent and, at worst, very actively antagonistic. Too often, moreover, these organizations are at the mercy of funders whose agendas range from protecting wetlands to keeping disposable diapers out of landfills.

"These groups are running around trying to put out all these fires," environmental journalist Dianne Dumanoski said, "but nobody's going after the pyromaniac."

Furthermore, a majority of climate groups—shackled by the cautiousness of their conservative funders and afraid of appearing too radical to the business and political establishments—continue to promote market-based approaches as a way to propel a global transition to clean energy.

The most popular of these approaches involves international emissions trading. The mechanism of emissions trading was developed by the environmental group Environmental Defense and championed by then vice president Al Gore in the mid-1990s. The rationale was that a system based on economic markets would help all players find the cheapest and most rational way to reduce carbon emissions. It assumes that nature will accommodate our economic system. It will not. There is no way that a market-based system can accomplish a global transition to clean energy.

There is an Alice in Wonderland quality in the thinking of many policy experts who regard the abstraction of a market economy as more real than the natural world that is imploding all around us. (A more detailed critique of the mechanism of the international "cap and trade" system appears in the next chapter.)

Amory Lovins, director of the Rocky Mountain Institute, has identified and developed an extraordinary array of efficiency technologies that, if adopted, would reduce our carbon emissions considerably. Whereas most economists estimate that the industrial world could cut its aggregate emissions by about 30 percent through efficiencies, Lovins contends that figure is closer to 50 percent. The ingenuity of the institute's design for a lightweight, safe, and amazingly efficient "hypercar"—to cite just one example—is truly revolutionary.

Unfortunately, Lovins's understanding of current economic dynamics seems off target. He assumes, since efficiencies save money, that new and reliable efficiency technologies will automatically be adopted by the marketplace. But that optimism seems unwarranted. The flaw lies in the assumption that our activities are governed by a free market. In truth, there is no free market in energy. Energy is probably the most highly regulated commodity there is—from price fixing by OPEC, to utility protection regulations, to the regulating of automobile mileage standards, to government subsidies for coal and oil and a host of other local, state, and federal regulations. Beyond that, the regulations that inform the production and distribution of energy are rigged to conform to the requirements of the coal, oil, and auto industries as they are currently configured. They provide little space for efficiency innovations and even less for renewable energy technologies.

Nevertheless, were the climate crisis confined to the industrial world, one could argue that markets, alone, might possibly be able to address it. But given the energy needs for the survival of the poorest two-thirds of humanity—and the inability of developing countries to finance new energy infrastructures—there is no alternative to a regime of international regulation. One such example, the World Energy Modernization Plan, is detailed in the conclusion of this book.

One cluster of climate organizations focuses on the corporate world—hoping that by enlisting high-profile, climate-conscious businesses to reduce their emissions, they will successfully promote the message that energy conservation not only enhances a company's public image but cuts operating expenses as well.

That message may be true for a small segment of companies whose production processes involve large amounts of energy use. For the vast majority of companies, however, energy costs amount to about 2 percent of their business expenses—hardly a large enough part of their bottom lines to require energy adjustments. But the argument is also disingenuous. Climate change will not be solved through energy efficiency. It requires an energy revolution.

The corporate focus that several groups have adopted, moreover, runs the risk of caricaturing itself. Business-oriented climate groups—just a handful of them—are all running after the same small group of companies that have evidenced some corporate concern about the climate.

For example, the Washington-based Pew Center for Climate Change, as noted above, has recruited about fifty highly visible companies to its Business Environmental Leadership Council. Most of these firms are implementing energy reductions in their internal operations. A number have agreed to support mandatory reporting of their carbon emissions. And they include some very high-profile corporations—BP, Alcoa, DuPont, Baxter International, Entergy, Royal/Dutch Shell, and Interface, Inc., among others.

These corporations, along with several others, have become the darlings of the climate movement. As a result, numerous environmental groups are all trotting out many of the same companies to trumpet their successes.

For example, Environmental Defense, one of the country's largest national environmental groups, boasts on its Web site: "With U.S. government policy in flux, we mobilized the business community. Entergy Corporation, one of the nation's largest utilities, joined Alcan, BP, DuPont, Ontario Power Generation, Pechiney, Shell International and Suncor Energy in our Partnership for Climate Action. The companies will report their greenhouse gas emissions publicly and each has set a firm target for reducing emissions. Entergy plans to hold carbon dioxide emissions constant even as it increases its non-nuclear electric generating capacity by about 28%. The New Orleans-based utility will improve its power plants, increase renewable energy capacity and invest in outside projects that reduce emissions."

Virtually all the same companies are featured in the Pew Center's business leadership council. Another Washington-based group, Cool Companies, praises BP for promising to cut greenhouse emissions 10 percent below 1990 levels, by 2010. It cites Royal Dutch/Shell as planning to cut emissions 25 percent by 2002. And it notes that DuPont pledged to reduce its greenhouse emissions 65 percent from 1990 levels by 2010; purchase 10 percent renewable energy by 2010; and hold energy use flat, using 1990 as a base year.

These companies surely deserve praise for voluntarily reducing their energy use. But none of them is moving beyond internal improvements into the public policy arena.

Seen through a different lens, the efforts of such business-oriented climate groups as the Pew Center, Environmental Defense, Cool Companies, and Clean Air–Cool Planet, a New England–based regional group, might be far more effective if they required their business partners to form aggressive lobbies in Washington to put real economic muscle behind corporate demands for radical changes in the country's energy policies.

The fossil fuel lobby has hijacked America's energy and climate policies. An appropriate response would seem to require a coalition of corporate and financial institutions of equivalent force and influence to counteract the carbon industry's stranglehold on Congress and the White House. But, at this point, that seems beyond the imagination or the capability of any of the corporate-focused climate groups.

A few organizations such as Campaign ExxonMobil, the Interfaith Center for Corporate Responsibility, and Greenpeace, USA have taken a more confrontational approach to pressuring big coal and oil firms. Their work paid off in a stunning victory at the 2002 annual meeting of ExxonMobil. After lobbying a number of institutional and activist shareholders, Campaign ExxonMobil managed to place a resolution on the agenda of the ExxonMobil annual meeting. In essence, the resolution called on ExxonMobil to stop its campaign of disinformation and develop its own plan for renewable energy sources.

Normally, dissident shareholder resolutions are considered successful if they marshal 5 or 6 percent of the vote. But in May 2002, the alternative resolution at the ExxonMobil annual meeting netted an extraordinary 21 percent. The vote sent shock waves through the corporate suites at ExxonMobil.

As a result of the shareholder campaign, the oil giant publicly softened its stance toward global warming—most visibly by toning down the rhetoric in its advertisements on the op-ed page of the *New York Times.*

The vast majority of climate groups, however, shun confrontation and work instead to get people to reduce their own personal energy footprints. That can certainly help spread awareness of the issue. On a superficial level, however, there is an aesthetic and emotional disconnect between the huge drama of climate change and the pedestrian nature of carpooling and compact fluorescent bulbs. The mismatch between the magnitude of the problem and the seductiveness of easy—and illusory—solutions reflects a degree of denial among even the most earnest advocates.

This approach may also conceal some counterproductive consequences. By persuading concerned citizens to cut back on their personal energy use, these groups are promoting the implicit message that climate change can be solved by individual resolve. It cannot.

By appealing to lifestyle changes, the climate groups may well be undermining a rising sense of anxiety and anger within the public. Rather than demanding political action on a national scale, many concerned citizens may come to feel they have made their contribution to the climate fight by carpooling, turning down their thermostats, and using less hot water.

The implicit message behind this approach is one of blaming the victim: People are made to feel guilty if they own a gas guzzler or live in a poorly insulated home. In fact, people should be outraged that the government does not require automakers to sell them cars that run on clean fuels, that building codes do not reduce heating and cooling energy requirements by 70 percent, and that government energy policies do not mandate decentralized home-based or regional sources of clean electricity.

What many groups offer their followers instead is the consolation of a personal sense of righteousness that comes from living one's life a bit more frugally. That feeling of righteousness, coincidentally, is largely reserved for wealthier people who can afford to exercise some control over their housing and transportation expenditures. Many poorer people—who cannot afford to trade in their 1990 gas guzzler for a shiny new Toyota Prius—are deprived by their circumstances of the chance to enjoy the same sense of righteousness, illusory though it may be.

Nevertheless, Web sites and flyers too numerous to mention instruct anxious members of the public to run their dishwashers only when full, to wash their clothes in cool or warm rather than hot water, to insulate their hot water heaters and keep their tires properly inflated—even as the world's glaciers are disappearing, whole islands are going under from rising sea levels, and world food supplies are dropping at alarming rates.

Still, despite their low goals and local focus, many activists are succeeding in raising the visibility of the climate issue all over the country. Given the antagonism of the Bush administration, the intense opposition of the carbon lobby, the difficulties in obtaining funding, and the unconscionable failure of the press to trumpet the urgency of the climate crisis, a number of local activists have achieved some real successes within the range of activity they deem possible.

One category of grassroots organizations around the country is reacting to the absence of federal leadership by pressuring city and state officials to begin to reduce carbon fuel emissions at the local level. While all these initiatives are laudable, none is really effective. Virtually all are geared toward reducing emissions domestically by 7 percent below 1990 levels, by 2012. Virtually none are guided by the findings of some of the world's leading energy specialists—that we need to be deriving half our energy from noncarbon sources by 2018.

Many groups are working with the International Council for Local Environmental Initiatives to reduce emissions in more than 110 cities across the country. Other groups are also working with some thirty states in developing or implementing emissions-reduction strategies.

In New England, climate groups are helping implement the most ambitious of these plans—a regional compact signed by the governors of the six New England states and the premiers of the five eastern Canadian provinces to roll back emissions to 1990 levels by 2010 and by 80 percent at some undefined date in the future.

The victories of some of these local efforts—given the inertia of the general public and the low level of expectations— are impressive:

- Local activists helped push the mayor of Seattle to commit that city to exceed the Kyoto targets.
- They spearheaded the approval by voters of a $100 million bond for wind and solar energy in San Francisco.
- In Texas, activists have pushed the legislature into mandating that the state will get up to 5 percent of its electricity from wind and other renewable sources by 2010.
- The Tufts University Climate Initiative, designed to bring the university's emissions within Kyoto limits, is being replicated at an estimated 400 colleges and universities around the country.

One small, shoestring organization based in Portland, Oregon, has generated a huge amount of climate activity all over the country. The Green House Network was founded in 1999 by Eban Goodstein, an economics professor at Lewis and Clark College. Goodstein and his colleague, Matthew Follett, organize three-day sessions for students, professors, artists, engineers, and religious activists from around the country, training the activists in speaking and organizing to mobilize support among the general public. In its three years of existence, the Green House Network has trained nearly 250 activists in thirty states who give talks in communities around the country. In two years, "graduates" of the network gave more than 600 presentations in thirty states. The network has also sponsored a series of successful runners' races—"Race to Stop Global Warming"—in eight cities.

The network has contributed to a remarkable proliferation of climate groups over the last decade. Visible groups are flourishing in such cities as Seattle, Minneapolis, Philadelphia, Boston, New York, Washington, Miami, Portland, Austin, and elsewhere. In the process, they are both informing and inspiring individuals, congregations, campus groups, and local organizations around the country to begin to take action on the climate crisis.

There is also a noticeable growth of climate activists within the religious community. Groups like Interfaith Power and Light (and their state-based counterparts) are mobilizing to buy green energy for churches. Increasing numbers of ministers and rabbis are addressing the moral dimensions of climate change in their sermons. Other faith-based organizations are working with their congregations to integrate the climate crisis into the area of religious education. One group, Religious Witness for the Earth, works with delegations to the United Nations to make this issue less a subject of diplomatic maneuvering and more a matter of our common instincts of ethics, stewardship, and morality.

The upshot of all this scattered activity is the emergence of an extremely dedicated and enthusiastic core of activists. Operating below the radar of the national media, these groups are, like an invisible army of termites, quietly under-mining the foundations of industry-manufactured denial.

The overriding justification for all this action lies in its ultimate potential for creating a truly meaningful move-ment. There are other benefits as well. On a psychological level, hands-on activism presents an antidote to the kind of paralysis that can result from truly apprehending the magni-tude of this issue. University work is particularly worthwhile because our educational institutions are the traditional seedbeds of social progress. We look to our colleges and uni-versities for new ideas and new directions. If students can make climate change a priority on their campuses, that mes-sage will begin to reverberate through the larger society.

But the greatest value of all these efforts lies not in the amount of emissions that are reduced. Those reductions are negligible compared to nature's requirement for a stable climate.

The true importance of the work is its contribution to the creation of a political base to counteract the immense power of big coal and big oil. Given the lock on Congress and the White House by the carbon lobby, there is no way the U.S. government will ever move toward a rapid global energy transition without a massive uprising of popular will.

One bit of good news is that in Washington, this issue has become far less politically polarized between liberals and con-servatives. Certain conservatives, including William F. Buckley Jr., Senator Richard Lugar (R–Indiana), Jim Woolsey (former head of the CIA), John McCain (R–AZ), and even Paul O'Neill (former secretary of the treasury), have spoken out about the profound dangers embedded in climate change.

The biggest challenge to the environmental community is to overcome its own tendencies toward exclusivity and turf protection—and to reach out to mobilize a sufficiently large and vocal political base to force action to head off runaway climate change.

But this may be bitter medicine for environmentalists. Climate change is no longer the exclusive franchise of environmental groups. Environmentalists should be forging urgent alliances with other activists who focus on international relief and development, campaign finance reform, corporate accountability, public health, labor, environmental justice, and human rights—not to mention the highly energized communities of faith—to mobilize a broad and inclusive constituency around the issue.

The challenge of organizing such a broad-based alliance is great, especially given the priorities of other groups who see their own goals as increasingly urgent. But, presented in its true dimensions, the climate crisis contains very relevant aspects for each of these constituencies.

Groups like Oxfam, for instance, which deal with relief efforts from extreme weather events, are also extremely interested in promoting clean energy in developing countries.

For campaign finance reformers, the climate crisis has become a prime example of the contamination of our political system by money. What we are seeing in the Bush administration's climate policies do not reflect political conservatism. They reflect corruption disguised as conservatism.

Corporate accountability groups could play a very particular role in focusing the spotlight on the most recalcitrant and obstructionist of the oil and coal companies. The behavior of big coal and big oil, their economic rationalizations notwithstanding, is an unconscionable crime against humanity.

For public health groups, the increase of inner-city asthma and respiratory distress coincides with both increased emissions and the proliferation of allergens from escalating carbon levels. The spread of West Nile virus and Lyme disease domestically and the spread of dengue fever and malaria abroad all are driven by climate change.

For labor, the production of renewable energy devices is far more labor intensive than coal and oil extraction, which are essentially capital-driven, automated operations. If a climate program guaranteed that the nation's 50,000 coal miners would be either bought out or trained for other jobs, the support of labor would surely follow. As Alden Meyer, an energy-policy specialist at the Union of Concerned Scientists, has pointed out, in every previous energy transition—from wood to coal, from coal to oil, from oil to gas—each new source was far cleaner and far more efficient that the preceding source. More to the point, each energy transition has generated an explosion of new jobs and overall economic growth.

On both a domestic and global level, climate change is nothing if not an issue of environmental justice and human rights. The rights to secure shelter, food, and the tools for basic sustenance are embedded in the Universal Declaration of Human Rights—and we are already seeing them disrupted by our indifference to an increasingly unstable climate, an indifference that the British medical journal *The Lancet* recently called "a form of bio-political terrorism."

Climate change hits poor countries hardest—not because nature discriminates against the poor but because poor countries cannot afford the kinds of infrastructure needed to buffer its impact.

Several years ago, Hurricane Mitch killed 11,000 people in Central America. No hurricane in the United States has ever taken that kind of toll.

In May 2003, the worst flooding in memory in Sri Lanka killed about 300 people, left another 500 missing, and made 350,000 people homeless.

Rising sea levels are forcing the evacuation of whole island nations. Two years ago, officials began planning the relocation of 1,000 residents of the Duke of York Islands near Papua New Guinea, which are being inundated by rising waters.

The linkage between environmental justice and the climate crisis was articulated eloquently by Ambassador Lionel Hurst of the island nation Antigua and Barbuda in the spring of 2003: "The most populous and wealthiest of the world face a moral challenge greater than colonialism or slavery. They are failing in that challenge. Men have lost reason in the fossil fuel economy . . . Inhabitants of small islands have not agreed [to be] sacrificial lambs on the altar of the wealth of the rich."

The tragedy underlying the failure of the activist community lies in the fact that so many talented, dedicated, and underpaid people are putting their hearts and lives on the line—in ways that will make little if any difference to the climate crisis. They are outspoken in their despair about what is happening to the planet. They are candid about their acceptance of a self-defeating kind of political realism that requires relentless accommodation. What is missing from virtually all these groups is an expression of the rage they all feel.

When a small, unimposing woman refused to yield her bus seat to a white man in Montgomery, Alabama, it led to more than some sympathetic shoulder shrugging. It led to a few brave African American students demanding service at a white-only diner. And that, in turn, led to a movement that refused to be stifled until it had achieved full voting rights, equal job opportunities, and a full and complete measure of political representation—with or without the approval of the majority of the country.

The United States, similarly, did not withdraw from Vietnam because a few individuals moved to Canada or Sweden to avoid military service—or because the leaders of the antiwar movement were able to negotiate a reduction in the number of bombing runs over North Vietnam. The United States left Vietnam because of a succession of massive demonstrations and dramatic episodes of civil disobedience that the press could not ignore. Ultimately, the United States withdrew from that war because of a sustained uprising of popular will that ultimately forced one president of the United States to drop his plans for reelection and pressured his successor to scramble until he had achieved something he could call "peace with honor."

These comparisons to the climate movement may be seen as too harsh until one considers the most fundamental fact about the climate crisis.

Activists compromise. Nature does not.

Snapshots of the Warming No. 6

Less spectacular, perhaps, than the specter of rampant disease and heat stress—but far more debilitating in the long run—is the looming threat of malnutrition, with all its attendant illnesses, that faces growing numbers of people around the world.

Warming temperatures, as well as altered rainfall patterns and more extreme droughts, are beginning to take a devastating toll on the world's food crops. For the first time in history, according to a report by the Earth Policy Institute, the world's grain harvest has fallen short of consumption levels for four years in a row.

"Higher temperatures are thwarting farmers' efforts to expand food production. The earth's average temperature has been rising since the late 1970s, with the three warmest years on record coming in the last five years. As temperatures continue to rise, crop yields start to fall.

"Last year India and the United States suffered sharp harvest reductions because of record temperatures and drought. This year Europe bore the brunt of higher temperatures. Record heat

in late summer scorched harvests from the United Kingdom and France in the west through Ukraine in the east. Bread prices are rising in several countries in the region," the report noted.

Chronicling the progressive decline in grain yields, the report noted that "in 2000, the shortfall was a modest 16 million tons; in 2001 it was 27 million tons; and in 2002 a record-smashing 96 million tons. In its September 2003 crop report, the U.S. Department of Agriculture reported that this year's shrunken harvest of only 1,818 million tons is falling short of estimated consumption of 1,911 million tons by a near-record 93 million tons."

The prognosis for the future of food production in the United States does not bode well. A new study by researchers at the National Center for Atmospheric Research, based on more geographically fine-tuned models, shows a few benefits—and much larger deficits—facing U.S. agricultural output.

Although corn production in the Northern Plains is projected to increase as temperatures rise, productivity will drop throughout the Midwestern Corn Belt. In the southeastern United States, researchers project a startling one-third loss to the agricultural economy if farmers don't prepare for climate change—and a one-fifth loss even if they do change crops to reflect warmer conditions. Cotton is expected to thrive under the warmer climate, but most food crops will become much more fragile and unreliable.

The projected crop declines are due largely to several factors. Warming will dry out soil more quickly, while altered rainfall patterns will result in significantly less rain falling in agricultural areas during the critical summer growing season, according to the research.

As the world's food supplies wither under increasingly warm and water-stressed conditions, increasing numbers of people will find themselves with compromised immune systems, increased infections, more extensive lethargy, and a host of disabling conditions. Some 12 million deaths were attributed to malnutrition in 1990, but that figure is expected to escalate dramatically as food supplies become progressively more scarce and more expensive. According to one report, 50 percent of women in developing countries are already anemic—and 3.6 billion people currently suffer from iron deficiencies.

As noted in a report by the World Resources Institute: "[M]alnutrition can impair the immune system, leaving malnourished children less able to battle common diseases such as measles, diarrhea, respiratory infections, tuberculosis, pertussis, and malaria. Vitamin A deficiencies are often worsened by infectious disease; and reciprocally, poor vitamin A status is likely to prolong or exacerbate the course of an illness such as measles. Similarly, malaria parasites, which require iron in order to multiply in blood, can cause or exacerbate anemia. Malnutrition can also heighten the adverse impacts of toxic substances. Deficiencies of protein and some minerals, for example, can significantly influence the absorption of lead and cadmium into the body."

The report noted that Africa and Southeast Asia confront problems of both malnutrition and such diseases as diarrhea, malaria, and measles—"a combination that is likely to increase the toll that either problem would take alone. In rapidly industrializing cities with high levels of malnutrition as well as disease and growing industrial pollution, residents may confront a triple burden of malnutrition, infection, and toxic pollution."

Ironically, the historically agriculture-rich country of Australia is also facing food problems—but for a very different reason. An eighteen-month drought in Australia has cut farm incomes in half and triggered fears of a permanent drought in the southwestern section of the country. Throughout history, farmers around the world have become accustomed to periodic droughts, floods, and weather aberrations. But Australians worry that this protracted lack of water in a major food-growing area may become a new and permanent condition.

The reason involves yet another obscure and largely unanticipated consequence of climate change. Both the North Pole and the South Pole sit at the center of a system of strong, swirling winds. At both poles, scientists have documented a tightening of the vortex of those polar winds. As the vortices tighten, the winds become stronger and tighter, shortening their range and intensifying their power. Scientists documented the increased velocity of the wind spin at the South Pole by comparing atmospheric pressures over the Antarctic continent to the same pressures over the southern ocean near Australia. While the polar area is cooling, the Australian landmass, further north—like most of the rest of the world—is warming.

In the north, where the landmasses of Europe and North America extend far inside the Arctic Circle, these winds are increasing the difference between temperature gradients over land areas and leading to much sharper and more heavily contrasting weather fronts—a phenomenon that some scientists associate with a marked increase of tornadoes in North America.

In Australia, however, where the landmass is farther away from the South Pole, the consequence is different. The strength-

*ening winds around the South Pole are drawing the water va-
por that rises from the South Pacific down toward Antarctica.
Water that normally rained down on Australian food-growing
areas is now being pulled farther south—leaving much of the
southwestern section of the island continent in drought. "There
has been real added impetus here in Australia to try to study
[the wind vortex] because we've been faced with an almost pre-
cipitous rainfall decline," said Dr. James Risbey of the Center
for Dynamic Meteorology and Oceanography at Melbourne's
Monash University.*

7

Thinking Big: Three Beginnings

To me the question of the environment is more ominous than that of peace and war . . . I'm more worried about global warming than I am of any major military conflict.
✍ U.N. WEAPONS INSPECTOR HANS BLIX
MARCH 14, 2003

Given the inadequacies of small-scale reductions, the obsolescence of the initial Kyoto goals, and the failure of unregulated markets to propel a global energy transition in time to avert severe climatic disruptions, it is worth looking at several large-gauge, intellectually courageous proposals that have been put forward in the last several years.

But before exploring these proposals—Contraction and Convergence (C&C), Sky Trust, and the Apollo Project—it is useful to view them in the context of the Kyoto process, which has guided the international effort to harness climate change for the last decade.

The initial response of the world's governments to news that the climate was warming was to draft the U.N. Framework Convention on Climate Change, which in turn gave rise to the Kyoto Protocol.

In response to concerns about the economic impact of an overly rapid transformation of national energy structures—and also about the ability of poor countries to afford a transition to clean energy—the delegates to Kyoto declared that the treaty's goal would require industrial nations to cut their aggregate carbon emissions by 5.2 percent below 1990 levels, by 2012. The developing countries would be obligated to cut their emissions in a subsequent round of the treaty.

The treaty exempted the developing countries from the first round of cuts—a policy that was affirmed by the first President Bush in 1992 when he signed the Rio Treaty. The reasoning for that exemption was clear. The industrial nations created the problem, and they have the resources to begin to address it. The industrial world needs to take the lead and the rest of the world will follow.

A central feature of the Kyoto Protocol was its reliance on a mechanism called "emissions trading." Under this mechanism, the nations of the world agree on an overall, total cap for carbon emissions. Then countries that exceed that cap can buy credits from countries that do not exceed it. For instance, the United States (or big carbon emitters within the United States) could pay to protect and expand forests in Central America to absorb some of the excess carbon dioxide emitted inside the United States.

In June 2003, the European Union established a formal "emissions trading" market, which is scheduled to take effect in 2005. Domestically, Governor George Pataki of New York announced a formal carbon trading market involving eleven states—to take effect in 2005.

Unfortunately, the mechanism of international emissions trading is dismally inadequate to the task on a number of counts.

On one level, most trading proposals involve reforesting areas in developing countries or protecting existing forests and grasslands to absorb carbon emissions. But the science underlying the reactions of vegetation to enhanced carbon concentrations seems not to have penetrated the policy arena. A slew of recent scientific findings have determined that the capacity of the world's forests and grasslands to absorb more carbon dioxide is about exhausted.

One team of seventeen scientists, studying the carbon-absorption potential of the world's vegetation, concluded in a report in the journal *Science* in 2000, "[T]here is no natural savior waiting to assimilate all the . . . CO_2 in the coming century."

Another team of scientists, headed by William Schlesinger of Duke University, surrounded pine trees with enhanced carbon dioxide for seven years. At the end of the experiment, Schlesinger concluded that the world cannot rely on forests to store our excess carbon dioxide. The best solution, he added, is to burn less coal, oil, and natural gas. "Rather than trying to gather up marbles that have spilled, let's not spill 'em in the first place," he said. As to the potential of the world's vegetation to absorb heat-trapping carbon, Schlesinger told *U.S. News & World Report:* "I would count on nothing."

But the flaws underlying the mechanism of international emissions trading involve more than plant biology. Emissions trading can work relatively well within nations. Domestic cap-and-trade programs—such as the U.S. trading program set up to reduce sulfur dioxide emissions—were relatively successful because they are easy to monitor and enforce. Most of the SO_2 emissions came from 2,000 smokestacks in the Midwest, a manageable number to monitor. The program, moreover, was subject to an enforceable system of national regulation.

But at the international level, the system of "cap-and-trade" totally breaks down. It is not monitorable. It is not enforceable. Moreover, it is plagued by irreconcilable equity disputes between the industrial and developing countries.

For one thing, there are far too many sources of carbon dioxide around the world to effectively monitor emissions—without turning half the world's population into carbon police.

At another level, there is a profound controversy between industrial and developing countries over how to allocate emission rights. The industrial nations want each country's emission rights based on its 1990 levels to ensure continuity of their economies.

Many developing countries, by contrast, contend that only a global per capita allocation is fair and democratic. Every inhabitant of the planet, they argue, should have equal rights to pollute the atmosphere. The problem here is that the typical American is responsible for twenty-five times more carbon emissions than the typical resident of India. So if the emission quota for each U.S. citizen were the same as for each citizen of India, that would decimate the U.S. economy.

A second equity issue, articulated by the late Anil Agarwal, founder of the Centre for Science and Environment in New Delhi, focuses on provisions in the Kyoto Protocol that allow industrial nations to buy limitless amounts of cheap emission reductions in poor countries and to bank them indefinitely into the future. This means that when developing nations eventually become obligated to cut their own emissions, they will be left with only the most expensive options since the cheaper offsets will have already been bought up by industrial countries. This clearly constitutes a form of environmental colonialism.

Finally, even if all the shortcomings involving monitoring, enforcement, and equity could be resolved, international carbon trading would most appropriately be used as a fine-tuning instrument—to help countries attain the final 10 to 15 percent of their obligations. This is not the workhorse vehicle we need to propel a worldwide energy transition.

We cannot finesse nature with accounting tricks.

In the absence of either a strong and binding Kyoto Protocol or an equivalent commitment by the United States, advocates have put forth a number of sweeping proposals, all of which are currently vying for the attention of policymakers, major foundations, and sympathetic labor and corporate organizations.

The major ones include Contraction and Convergence, Sky Trust, and the Apollo Project. Although each of these programs emphasizes different mechanisms, they are united by a common goal—a rapid transition to noncarbon energy. Unfortunately, each seems flawed in its own way. And all of them seem to ignore the extraordinary transformative potential—politically, diplomatically, and economically—that the crisis offers.

Nevertheless, these proposals reflect ingenious and innovative programs that have been developed by a number of very thoughtful and well-informed people. Given the reach of these proposals—and the understanding by their authors of the real dimensions of the climate crisis—they all may contain elements that could contribute to a truly effective, comprehensive approach.

ℒ♥

The Contraction and Convergence model was developed about ten years ago by Aubrey Meyer in Britain. Meyer, a member of Britain's Green Party, cofounded the Global Commons Institute in London in 1990. Since then, the work of the institute has centered around promoting C&C to address the shortcomings of the Kyoto Protocol.

At the heart of the plan is an elegantly simple scheme whose primary goal is the assurance of equity for developing countries in the global push to avoid climate chaos.

Under the plan, the governments of the world would gradually begin to reduce emissions on a basis that is roughly proportional to their populations. If C&C were adopted as the tool for managing CO_2 and other greenhouse gases, "there would be a transition to a point (convergence) where future entitlements to emit will have become proportional to population," according to Meyer. In other words, by the year 2045—which is a hypothetical target date set by the institute for the world to stabilize atmospheric carbon concentrations—each resident of the planet would be allowed to emit (or be responsible for) the same amount of carbon dioxide as everyone else. The plan would basically "piggyback" on a regime of international carbon trading so that by the time the trading goals were met, every resident of the planet would enjoy the same allotment of carbon energy.

The C&C plan is driven by the major concern that the industrial nations, in particular the United States, will devise a way of bringing down the world's aggregate carbon emissions and, at the same time, either perpetuate—or, more likely, intensify—the relative poverty of the developing world. In other words, the countries of the North will try to achieve climate stability on the backs of the world's poor.

The premise is surely justified historically. The shortcoming lies in the reliance of C&C on a prolonged regime of "emissions trading" to redress the kind of inequities that mark relations between industrial and developing countries. If the world moves past the very preliminary stages of addressing the climate crisis, it will abandon emissions trading as an inadequate tool—and, with it, the promise of global equity that sits at the core of C&C.

Ultimately, the C&C system lacks a mechanism to determine either a mandatory time frame or a mandatory limit of future carbon emissions. In its model scenario, the Global Commons Institute assumes a hypothetical date of 2045—and a concentration of carbon levels at 450 parts per million. But if the world were to wait until 2045 to bring concentrations down to 450 parts per million, it seems highly likely that the climate crisis will already have spiraled out of control. As cited earlier, a team of researchers, writing in both the journals *Nature* and *Science* have indicated that the world needs to be generating half its energy from noncarbon sources by 2018 to avoid an inevitable doubling of preindustrial carbon levels (at 560 ppm) and even a possible tripling of those levels—either of which would clearly be catastrophic. It is worth repeating that we are already witnessing major, perhaps irreversible, climate disruptions at the current level of carbon concentration of 370 ppm.

The proponents of C&C—who include Meyer as well as Tom Athanasiou and Paul Baer, who coauthored the 2002 book *Dead Heat*—espouse a rights-based, equity-oriented approach to climate stabilization that appeals especially to people concerned with economic and environmental justice. Athanasiou and Baer proposed a more finely tuned version of contraction and convergence that accounts for different geographical conditions and historical factors, especially in developing countries.

But the overall approach lacks both the economic incentives and binding regulatory mechanisms to force the kind of transition that is needed. To put the critique in its harshest terms, a regime of contraction and convergence, as currently conceived, would, at best, ensure that energy cutbacks are more equitably imposed even as the world moves forward into an increasingly destructive new climate regime.

⚜

A second large-scale proposal currently under consideration by the climate policy community is called Sky Trust. The plan was conceived by Peter Barnes, a cofounder of Working Assets Long Distance, who has long been preoccupied by the failure of will in the United States to address the coming climate crisis.

In his book *Who Owns the Sky?* Barnes modeled his carbon reduction program after the Alaska Permanent Fund, which was developed by former governor Jay Hammond of Alaska. Hammond realized in the late 1970s that the state of Alaska was about to enjoy a major windfall with the advent of large-scale oil development on Alaska's North Slope. His question was how to use that windfall for the maximum advantage of the state's residents. His efforts resulted in the creation of a permanent fund that was financed by 25 percent of the oil revenues the state received from oil exploration and drilling companies. The fund was then allocated equally among all the state's citizens—with every resident receiving one share of the fund.

In its first year of operation, the Alaska Permanent Fund wrote checks of $1,000 to each of the state's residents. By the year 2000, the payments had nearly doubled, to $1,964 per year per citizen.

Using the Alaska Permanent fund as a model, Barnes developed his Sky Trust plan. Since the sky is the ultimate repository of carbon dioxide, Barnes has proposed putting the sky into a trust—and dividing that trust equally among all U.S. citizens. Under his plan, every citizen of the United States would inherit one share of the sky. Those shares would then be priced to hold down emission levels by increasing costs to carbon emitters—even as it rewarded non- or low-polluting people with shares of the trust's proceeds.

Like the Contraction and Convergence model, Sky Trust basically capitalizes on the mechanism of emissions trading. Under Barnes's scenario, the trust would administer the rights to emit 1.3 billion tons of carbon per year—the same amount that the United States emitted in 1990, the baseline year used by industrial nations in the Kyoto Protocol.

While the price of those emission rights would ultimately be set by a trading market, an arbitrary initial permit price would not exceed $25 per ton. As Barnes pointed out in his book, a $25 per ton price is actually quite cheap—translating into the equivalent of about an extra 6 cents per gallon of gasoline. But since commodity prices increase in direct proportion to their scarcity, the permit costs would rise as more and more emitters competed for a finite number of carbon pollution rights. Under his plan, for example, cars that use more gasoline per mile would cost more because of permit requirements. And the number of permits would be decided each year by Congress.

The other side of the Sky Trust equation involves the repayment of the permit proceeds to all U.S. citizens, similar to the way in which the Alaska oil fund is spent. Every year, each citizen of the United States would receive a dividend from the trust. In one example, Barnes postulated that in 2010, each U.S. resident would receive $644 from the U.S. Sky Trust.

Sky Trust provides a model of transparency and accountability. It is economically symmetrical because it is funded by the appropriate offending activity (coal and oil burning)—and because it follows a fundamental market principle. In particular, as emission rights become progressively more scarce, their costs rise accordingly and their dividends increase apace.

Ultimately, I believe, it suffers from several shortcomings. First of all, although Sky Trust would begin to increase market-based emission costs within the United States, it contains no mechanism to bring down emissions from countries like India and China—both of which will shortly surpass the United States as the world's largest carbon emitters. It is exclusively a domestic program.

Nevertheless, as Barnes wrote in his book: "Our denial can't last forever." And when the United States finally does decide to cut emissions in a meaningful way, Sky Trust seems like a highly promising way to make those reductions far more equitable through its dividend program—especially since the poorest members of the U.S. economy are also its lowest emitters.

But like C&C, Sky Trust is not equipped to establish an ultimate goal for global carbon concentrations. Nor does it contain a mechanism to begin to roll back our fossil fuel usage at the rate required by nature to keep this planet hospitable to civilization.

꙳

The most recent large-scale attempt to move us off our track toward climate chaos is a recently born alliance between environmental groups and, at last count, about a dozen labor unions. The Apollo Project was established by the Institute for America's Future, the Center on Wisconsin Strategy, the Common Assets Defense Fund, Americans for Energy Freedom, and the Carol/Trevelyan Strategy Group.

Spearheaded by Bracken Hendricks, a former NASA worker who heads the Institute for America's Future; Michael Shellenberger, head of Lumina Strategies; and Adam Wehrbach of the Sierra Club, the Apollo alliance is geared toward having an impact on the 2004 presidential election. The campaign focuses primarily on creating what it estimates would be about 1 million new jobs in renewable energy from an Apollo Project. It promises to reduce and, ultimately, eliminate the U.S. dependence on foreign oil, which has created excruciatingly difficult diplomatic entanglements, has put the United States in the awkward position of supporting some of the world's most undemocratic regimes, and has made this country vulnerable to unpredictable developments in the politically volatile Middle East.

At its core, the Apollo Project is simple: It calls for the expenditure of $300 billion over a ten-year period on energy efficiency and renewable energy initiatives. The project's ten-point plan calls for that money to be spent on an eminently sensible array of projects, including, among other things, promoting hybrid and fuel-cell vehicles, encouraging more "green building" construction, expanding transportation options—more bicycle paths, better public transportation, and regional high-speed rail systems—and increasing investments in "smart growth" development that preserves open spaces, creates more pedestrian areas, and increases residential density.

Drawing on the public relations talents of its promoters, the Apollo Project gained visibility even before the 2004 presidential campaign was fully under way. A number of its proposals were embraced by Senator John Kerry (D–MA) and Governor Howard Dean (D–VT), both early front-runners for the Democratic presidential nomination. It also resurrected a fragile alliance between environmentalists and labor unions—after earlier such efforts had crashed and burned.

Unfortunately, from the point of view of the climate crisis, it shares a shortcoming with the Sky Trust proposal in that it is confined to the United States. There is nothing in the Apollo program that would deal with the spiraling levels of carbon emissions from developing countries.

Even more unfortunate from a foreign policy point of view, the Apollo Project's emphasis on energy independence could exert a very destabilizing impact on international relations. Were the project to be realized—and the United States to substantially reduce its purchases of foreign oil in a short time—it would rapidly drain away wealth from the oil-producing nations, especially in the Middle East. With nothing in the plan to accommodate Saudi Arabia, Kuwait,

Egypt, Iraq, and all the other oil-exporting nations, a successful Apollo Project could very well mark the beginning of a tumultuous and destructive wave of developments in the Middle East. It could easily exacerbate the perceptions, in much of the world, that the United States is antagonistic to the Muslim world. It would increase the attractiveness of anti-U.S. terrorism by intensifying poverty in areas that have been able to maintain domestic stability through the wealth of oil exports.

Finally, because of the delicate nature of forging an alliance between environmentalists and labor unions, the developers of the Apollo Project have chosen to promote the single largest contributor to global warming—our burning of coal. Coal emits about 30 percent more carbon per unit of energy than oil and about twice as much as natural gas. There is no way to address the climate crisis without very quickly eliminating coal from the world's energy diet. Unfortunately, the Apollo Project contains no mechanism to help transform the world's major fossil fuel producers into developers of renewable energy sources.

(It seems that compromise could have been avoided by the inclusion in the Apollo Project of a fund to retrain or buy out the nation's 50,000 coal miners—perhaps by siting some wind-manufacturing facilities in coal-rich areas. The real resistance would likely not come from the miners. Building windmills and solar panels is a much healthier way to make a living. The real resistance would come from their union leadership, which is concerned more about the survival of the union, itself, than about preparing its members for jobs in a new energy economy. As a result, the union leadership basically acts as an arm of the nation's coal corporations when it comes to the climate issue.)

Because of its politically motivated inclusion of coal, the Apollo Project also regrettably calls for huge investments in mechanical carbon sequestration. Under this strategy, carbon dioxide would be captured from coal-burning power plants and piped into burial areas deep inside mountains or mine shafts. It is an extremely risky and unproved method of reducing atmospheric CO_2. More to the point, it is extraordinarily wasteful. Even putting aside, for a moment, the question of their efficacy, the cost of a system of carbon sequestration plants amounts to economic lunacy. In this particular proposal, the goals of the Apollo Project and ExxonMobil are virtually identical.

Unfortunately, deep-mine burial of carbon dioxide is a very uncertain and expensive technology that could lead to massive releases of CO_2 in the future. It could also contaminate aquifers, increase the likelihood of earthquakes, and damage biological communities within storage sites, according to the Union of Concerned Scientists (UCS). Even under the most optimistic economic scenarios, a system of mechanical sequestration to capture and bury carbon dioxide from power plants would nearly double the costs of electricity to consumers, according to the UCS analysis. While it grudgingly endorses more research into mechanical sequestration, the UCS has concluded: "In no way should geologic carbon sequestration be seen as a 'silver bullet' to reduce emissions, nor should it be researched and developed at the expense of other environmentally sound, technologically feasible, and economically affordable solutions to climate change. Instead, the highest priority should go to technologies and policies that prevent emissions to the atmosphere in the first place, e.g., improving energy efficiency in power generation, transportation and buildings, developing renewable energy, and protecting threatened forests."

If the Apollo Project were to gain widespread political traction, it could play a role in influencing the outcome of the 2004 presidential election. But even if it succeeded beyond the wildest dreams of its adherents, it would not stop global warming.

Of all the proposals outlined above, none would really limit the political stranglehold of the fossil fuel lobby on our political system—and none would propel the industry into the kind of transformation that a sustainable future truly requires.

None presents the kind of regulatory regime that allows whole industries to move in lockstep toward a clean-energy future. As John Browne, CEO of BP, has pointed out, this change simply cannot happen through the conscientious efforts of a few individual companies. Real change requires that the governments of the world regulate the transportation and energy industries so that they can make the transition into a new regime in lockstep—without any company sacrificing its competitive standing within its own industry. Anything short of that will not work.

Finally, most of the approaches currently on the table are overlooking an enormous potential: The climate crisis offers us an extraordinary and historically unique opportunity to begin to reverse a set of very destructive global dynamics—political, economic, and environmental. Although proposals like Contraction and Convergence, Sky Trust, and the Apollo Project would, if successful, begin to reverse at least the contribution of industrial countries to the catastrophic climb in global carbon emissions, their narrow focus on emissions would squander an enormously transformative and truly revolutionary opportunity embodied in a coordinated global crash program to rewire the planet.

Energy is the nervous system of civilization. A globally coordinated, properly framed reconstitution of the global energy infrastructure contains the seeds for a wealthier, more democratic, and ultimately more peaceful world.

Any approach that does not contain that focus is turning its back on an opportunity for historical redemption that, given the escalating pace of climate change, may never again present itself.

Snapshots of the Warming No. 7

Beyond its negative effects on human health, some of the most graphic impacts of climate change—sea level rise, rising temperatures, and the increase in violent weather—are imposing heartbreaking changes on people's cultures, on their traditional sense of place, and even on ancient systems of belief.

Begin with rapidly rising sea levels that are threatening entire populations of island nations.

In November 2000, officials began the permanent evacuation of more than 40,000 people from their traditional home. As the British newspaper the Independent *noted, the evacuation of the first 1,000 residents of the Duke of York Islands off Papua New Guinea "could be a dress rehearsal for millions of people around the globe affected by rising sea levels."*

The islands, together with neighboring atolls such as Takuu, home to a small community of "singing" Polynesians, are likely to be the first to be engulfed by the effects of global warming. The islands are just 12 feet above sea level, and water levels are rising by 11.8 inches per year.

A similar evacuation has begun in the island nation of Tuvalu, where 10,000 people are in the process of relocating to higher ground.

As Leo Falcam, president of the Federated States of Micronesia, cautioned: The Pacific Islanders' "early experience with real consequences of global warming has been considered analogous to the canary in the coal mine—providing an early warning to the global community of its own impending doom."

Nor are Pacific islands the only casualties. Small island nations from Jamaica to the Philippines are threatened by rising sea levels. And according to scientific estimates, a rise of one meter—which has been predicted by the IPCC—would submerge 17.5 percent of Bangladesh, 6 percent of the Netherlands, and 80 percent of Atoll Majuro of the Marshall Islands.

The increasing temperatures in the atmosphere will also increase flooding, storm intensity, and the power of storm surges in areas that are especially vulnerable to such disruptions. As a result, the president of New Zealand has offered to take in Tuvalu's entire population. As the Honolulu Star *reported: "Climate change is not a future concern; it is an immediate threat. A diaspora has begun. The Tuvaluan people must build new lives in a new land."*

President Falcam was more succinct: "Climate change is nothing less than a form of slow death."

If climate change represents "slow death" to inhabitants of island nations, it represents a real-time nightmare to people living in the far North.

Warming is most pronounced at the higher latitudes. And nowhere have the combined impacts of global warming con-

verged more visibly than in Alaska. As Andrew Revkin of the New York Times *reported in June 2002:*

"To live in Alaska when the average temperature has risen about seven degrees over the last 30 years means learning to cope with a landscape that can sink, catch fire or break apart in the turn of a season. In the village of Shishmaref, on the Chukchi Sea just south of the Arctic Circle, it means high water eating away so many houses and buildings that people will vote next month on moving the entire village inland.

"In Barrow, the northernmost city in North America, it means coping with mosquitoes in a place where they once were nonexistent, and rescuing hunters trapped on breakaway ice at a time of year when such things once were unheard of. From Fairbanks to the north, where wildfires have been burning off and on since mid-May, it means living with hydraulic jacks to keep houses from slouching and buckling on foundations that used to be frozen all year. Permafrost, they say, is no longer permanent.

"On the Kenai Peninsula, a recreation wonderland a few hours' drive from Anchorage, it means living in a four-million-acre spruce forest that has been killed by beetles, the largest loss of trees to insects ever recorded in North America, federal officials say. Government scientists tied the event to rising temperatures, which allow the beetles to reproduce at twice their normal rate.

"In Alaska, rising temperatures . . . are not a topic of debate or an abstraction. Mean temperatures have risen by 5 degrees in summer and 10 degrees in winter since the 1970's."

At the other end of the world, in the African nation of Lesotho, the spring of 2002 brought with it climate chaos. First, there were unseasonably early rains. Then untimely frost. Then severe storms destroyed crops and created unusual famine conditions.

One local farmer summed it up: "Frost in the summertime! We never used to see weather like this. We don't know what to expect anymore from the skies. I think God is angry at us, but I don't know why."

8

Rx for a Planetary Fever

Our house is burning down and we're blind to it . . . The earth
and humankind are in danger and we are all responsible. It is
time to open our eyes. Alarms are sounding across all the conti-
nents . . . We cannot say that we did not know! Climate warming is still reversible. Heavy would be the responsibility of those
who refused to fight it.
☞ FRENCH PRESIDENT JACQUES CHIRAC
SEPTEMBER 2, 2002

It is hard to get people focused on climate change today,
not only because it has been misrepresented as a future
problem but because there is so much competition from
other problems.

We are under attack from terrorists. We are apprehensive
about the aftermath of the war with Iraq. The recent period
of economic stagnation has stunned us with the realization
that the global economy is just as vulnerable to abrupt and
unpredictable shocks as the nation's electricity grid. The

press has stubbornly refused to accord this story the attention and energy it requires.

So it is worth repeating that climate change is not just another issue in this complicated world of proliferating issues. It is *the* issue that, unchecked, will swamp all other issues.

Conversely, the solution to the climate crisis may well contain the seeds for solutions to some of the most threatening problems facing humanity today. The solutions to climate change have the potential to begin to mend a profoundly fractured world.

Take, for example, our newfound vulnerability to terrorism.

The most obvious connection is that the solution to the climate crisis—a worldwide transition to renewable energy—would dramatically reduce the significance of oil, and with it our exposure to the political volatility in the Middle East.

A second connection is that a renewable-energy economy would have far more independent sources of power—home-based fuel cells, stand-alone solar systems, regional wind farms—which would make the nation's electricity grid a far less strategic target for future guerrilla attacks. (Even absent terrorism, the vulnerability of large grid-based systems was underscored by the blackout of much of the northeastern United States in the summer of 2003. That power outage, whose reach exceeded a similar blackout in 1977, was the fourth such failure in the last decade, according to the Rocky Mountain Institute.)

More relevant to our security is the fact that poor countries are much more immediately vulnerable to the impacts of climate change. The continuing indifference by the United States to atmospheric warming—since this country generates one-fourth of the world's emissions with 5 percent of its people—will almost guarantee more anti-U.S. attacks from people whose crops are destroyed by weather extremes, whose populations are afflicted by epidemics of infectious disease, and whose borders are overrun by environmental refugees.

The real truth about terrorism is that, aside from hardening specific targets like airports and nuclear plants, there is no way to protect any complex organized society from guerrilla attacks. In the long run, what is really required is a major change in our posture toward developing countries.

Dr. Rajendra Pachauri, the new chairman of the Intergovernmental Panel on Climate Change, pointed out at the end of 2002 that the impacts of climate change "will exacerbate world poverty and could make millions of people more open to extremism." In an interview with Reuters News Service, Pachauri added: "Large areas of poverty are dangerous for the world as a whole as they provide fertile ground for extremist views. Things go wrong. People want to blame someone." In this case, the United States really is the major culprit.

Just as runaway carbon concentrations are threatening to destabilize the global climate, runaway economic inequity can only continue to destabilize our global political environment.

The prevailing view of developing countries as economic competitors reflects a near-mystical, almost fundamentalist, belief in the divinity of free markets—and an equivalent hypocrisy about the unacknowledged political control that industrial nations exert over those markets. In fact, the continuing economic exploitation of the developing world—through rigged trade rules, unfair subsidies in industrial countries, and the stultifying burden of chronic indebtedness—will only further impoverish poor countries relative to the industrial world. In shrinking foreign markets and preventing the sound development of an industrial base in many developing countries, this posture is basically a recipe for the slow-motion suicide of global capitalism.

Stepping back, it is worth repeating that the real economic issue in rewiring the globe with clean energy is not cost. The real economic issue is whether the world has a big enough labor force to accomplish the task in time to meet nature's deadline.

A properly funded global transition to clean energy would create millions of jobs in poor countries and substantially raise living standards in the developing world.

It is an article of faith among development economists that energy investments in poor countries create far more wealth and jobs than investments in any other sector. Were the United States to spearhead a wholesale transfer of clean energy to developing countries, that would do more than anything else in the long term to address the economic desperation that underlies anti-U.S. sentiment.

Beyond the threat of terrorism, the competition for the world's dwindling supply of oil is certain to be a major source of potential military conflict in the coming decades. "Many resources are needed to sustain a modern industrialized society, but only those that are viewed as being vital to national security are likely to provoke the use of military force when access . . . is placed in jeopardy. There is no question that oil [enjoys] this distinctive status," according to Michael Klare, a leading expert on international security issues.

On the economic front, it seems clear that the entire global economy is susceptible to periods of stagnation, even recession. In mid-2003, economists at the Federal Reserve were raising concerns about that most frightening of economic dynamics—deflation. A truly floundering economy seems relatively immune to tax cuts and interest rate reductions. A more proven remedy for long-term economic stagnation seems to lie in public works programs. Depression-era New Deal programs, such as the Works Progress Administration, implemented by President Franklin D. Roosevelt, created jobs for millions of unemployed people. A generation later, President Dwight Eisenhower's construction of a national interstate highway system—at the time the biggest public works program in history—provided a substantial number of government-funded job opportunities.

Today, a global public works program to rewire the globe with clean energy would be the most productive investment we could make in our future. Within a decade, it would begin to generate a major and continuing worldwide economic liftoff.

Given the ferocious resistance of big coal and big oil, it is clear that the economic interests of the fossil fuel industry are, in many ways, diametrically opposed to the stability and vitality of the larger economy.

For example, there are a large number of multinational corporations—many of them based in the United States—that have saturated their markets in the United States, Europe, and Japan. A large class of companies like Boeing, Gillette, Coca-Cola, Procter & Gamble, and many others see virtually all their future earnings growth coming from developing countries.

But as we have seen, the impacts of climate change hit poor countries first and hardest—not because nature discriminates against the poor but because most developing countries cannot afford the infrastructures to accommodate flood, droughts, severe storms, epidemics of disease, and incursions of environmental refugees.

The majority of the extreme weather events that have been rocking the world for the last few years have taken increasingly severe economic tolls on developing countries. That trend will only continue. And as it does, those global corporations that look to developing countries for their future profits will see that projected growth evaporate as climate disasters drain national budgets and shrink purchasing power in the developing world.

It is also becoming increasingly clear that climatic instability is beginning to threaten the keystones of the world's financial structure—its banks and insurers.

In the 1980s, the world's property insurers lost an average of $2 billion a year to weather extremes. In the 1990s, they lost an average of $12 billion a year. In 1998 alone, property insurers lost $89 billion to extreme weather events, which is more than they lost in the entire decade of the 1980s.

As noted earlier, the community of property reinsurers estimates that losses from climate impacts to the global economy will jump to about $150 billion a year within this decade, to $300 billion a year in the next few decades, and even to bankrupting the global economy by 2065.

By contrast, the dramatic expansion of the overall wealth in the global economy from a worldwide energy transition would create new markets and significantly invigorate existing ones.

There is, however, another line of argument that centers not on the economy but on the most basic human moral principles. When policymakers discuss climate change, they discuss it in economic terms—the costs of climate impacts versus the costs of addressing the issue. That is the vocabulary that counts today with many decisionmakers.

But, at root, climate change is not an economic issue. It is, first and foremost, a moral issue.

It is telling that of all sectors of civil society in the United States, perhaps the most proactive and energetic response to the climate crisis is coming from the religious community. Congregations around the country are mobilizing both practically and in the pulpits. A number of congregations are beginning to bulk-buy renewable energy for their churches, synagogues, and mosques. But even more critical than the buying power of churches is the resonance of unified voices of faith.

To continue to ignore climate impacts means putting at risk billions of poor people around the world who are much more immediately vulnerable to its impacts. It means

dishonoring all the effort and sacrifice of all those generations who have worked so hard to create this civilization we
enjoy today. Ultimately, it means consigning our children
to a future of chaos and disintegration.

What is really missing from the climate debate is an insistence on the moral imperative of truly facing this challenge in
all its dimensions. In this case, that moral imperative requires
an equivalent policy imperative.

One model of such a policy centers on a set of macro-
level, global-scale policies that would address both the extraordinary threat of climate impacts and the economic
desperation of the majority of the world's citizens.

The World Energy Modernization Plan contains three
interactive, mutually reinforcing strategies that are designed
to reduce carbon emissions by the 70 percent required by nature—at the same time as they would create millions of jobs
around the world, especially in developing countries.

The plan was developed in 1998 by an ad hoc, informal
group of about fifteen energy company presidents, economists, energy policy experts, and others (including the author), who met at the Center for Health and the Global
Environment at Harvard Medical School. Since that time,
the plan has been presented at side conferences to the climate negotiations in Buenos Aires and Bonn. It has been endorsed by a number of developing country nongovernmental
organizations (NGOs). It received a very positive reception
from the former CEO of Shell/UK, who was also director of
a G-8 Task Force on Renewable Energy, and it has attracted
the interest of a small number of senators and congressmen.

Most recently, it was endorsed by a former British ambassador to the United Nations.

To set the plan in its starkest context: The deep oceans are warming, the tundra is thawing, the glaciers are melting, infectious diseases are migrating, and the timing of the seasons has changed. And all that has resulted from *one degree* of warming. By contrast, the earth will become 3°F to 10°F warmer later in this century, according to the IPCC.

That background informs this set of strategies. These ideas have been vetted by a number of economists and energy policy experts, but they are still provisional. Some elements may require major surgery. Although its developers happen to think this proposal is elegant, they are not dogmatic about its particulars.

What they do believe—very strongly—is that these strategies present a model of the scope and scale of action that is appropriate to the magnitude of the climate crisis. And, to date, no other policy recommendations adequately address the urgency of the problem.

As Britain's prestigious Institute for Public Policy Research (IPPR) declared recently, "Kyoto will not stop climate change. The next international climate change negotiations must agree on a safe level of emissions in the long term and fair shares between nations." Tony Grayling of the IPPR added: "Future international climate change policy should be based on sound science and social justice, not the horse trading that characterized the negotiations for the Kyoto Protocol."

Those sentiments were echoed by Graeme Pearman, chief scientist at the Australian Commonwealth Scientific and Industrial Research Organization, who added: "Slowing the rate of emissions of carbon dioxide into the atmosphere will not stop the increase of its concentration and thus climate change. . . . [R]eductions of 70% or more in current global emissions are necessary in order to stabilize concentrations."

Because of U.S. recalcitrance, coupled with the escalating pace of climate change, the Kyoto goals (but *not* the Kyoto process) are today irrelevant. It is time to go straight for a global reduction of 70 percent. The hope is to get ideas of this scope into the conversation to help move it to an appropriate level.

The plan involves three interacting strategies:

- a change of energy subsidy policies in industrial countries
- the creation of a large fund to transfer renewable energy technologies to developing countries
- the subordination within a Kyoto-type framework of the mechanism of international emissions trading to a new model, a progressively more stringent Fossil Fuel Efficiency Standard, which rises by 5 percent per year

Although each of these strategies can be viewed as a stand-alone policy, they are better understood as a systemic set of interactive policies that could speed the energy transition far more rapidly than if they were implemented in piecemeal fashion.

On the issue of subsidies, the United States currently spends more than $20 billion a year to subsidize fossil fuels. Industrial-country subsidies for fossil fuels have been estimated at $200 billion a year. (That figure does not include another $15 billion in U.S. subsidies that is frequently cited by other economists. That figure represents the amount of military expenditures the United States spends to ensure the security of transportation of oil from the Middle East.)

In the industrial countries, those subsidies would be withdrawn from fossil fuels and equivalent subsidies would be established to promote the development of clean energy sources. Clearly a small portion of the U.S. subsidies must be used to retrain or buy out the nation's 50,000 or so coal miners. Their welfare cannot be sacrificed to the interest of climate stabilization. But the lion's share of the subsidies would still be available for use by the major oil companies to retrain their workers and retool to become aggressive developers of fuel cells, wind farms, and solar systems. In other words, the subsidy switch is intended as a tool to help oil companies transform themselves into renewable energy companies.

These strategies would best not be implemented in isolation. If the subsidy switch alone were implemented in industrial nations, it would promote the growth of the renewable energy industry in the North.

But, as we know, the problem is global in scope. Even if the countries of the North were to dramatically reduce emissions, those cuts would be overwhelmed by emissions from the large developing countries. Therefore, the second element of the plan involves the creation of a new $300-billion-a-year fund to help transfer renewable energy resources to poor countries. Virtually all poor countries would love to go solar; virtually none can afford it. The most air-polluted cities in the world today are in China, Mexico, Thailand, Chile, and other developing and transitional countries.

One attractive source of revenue to fund the transfer lies in a "Tobin tax" on international currency transactions, named after its developer, Nobel Prize–winning economist Dr. James Tobin. Tobin conceived of his tax, which has yet to be implemented, as a way to dampen the volatility in capital markets by discouraging short-term trading and encouraging longer-term capital investments. But it would also generate enormous revenues. Today the commerce in currency swaps by banks and speculators amounts to $1.5 trillion per day. A tax of one-fourth cent on a dollar would net about $300 billion a year for wind farms in India, fuel-cell factories in South Africa, solar assemblies in El Salvador, and vast solar-powered hydrogen farms in the Middle East.

Since currency transactions are electronically tracked by the private banking system, the need for a large new bureaucracy could be avoided simply by paying the banks a fee to administer the fund. That administrative fee would, to some extent, offset the banks' loss of income from the contraction in currency trading that would result from the imposition of the tax.

The only new bureaucracy envisioned would be an international auditing agency to monitor transactions to ensure equal access for all energy vendors and to minimize corruption in recipient countries.

Several developing-country commentators have suggested that corruption could be further curtailed by requiring recipient governments to include representatives of indigenous minorities, universities, NGOs, and labor unions in making decisions about the procurement and deployment of new energy resources.

If a Tobin tax proves unacceptable—and some economists express a nervousness about this untested mechanism—a tax on airline travel or a carbon tax in industrial countries, though more regressive, could fulfill the same function. Florentin Krause, of the IPCC's Working Group III, and Stephen DeCanio, former staff economist for the Reagan Council of Economic Advisers, estimate that if carbon emissions were taxed at the rate of $50 a ton, the revenue would approximate the $300 billion from a tax on currency transactions. (It is unclear what would happen to transitional prices of carbon fuels if subsidies were removed *and* a carbon tax imposed at the same time. That may, or may not, be an economically viable step.)

Regardless of its revenue source, the fund—on the ground—would be allocated according to a U.N. formula based on climate, energy use, population, economic growth rates, and so on to determine what percentage of each year's fund would go to each developing country.

If India, for instance, were to receive $5 billion in the first year, it would then decide what mix of wind farms, village solar installations, fuel cell generators, and bio-gas facilities it wanted. The Indian government (in this hypothetical example) would then entertain bids for the construction of new wind farms, solar panels, fuel cells, and other clean-energy sources. As contractors reached specified development and construction benchmarks, they would then be paid directly by the banks. And the banks, as noted, would receive a fee for administering the fund.

As self-replicating renewable infrastructures took root in developing countries, the fund could simply be phased out. Alternatively, progressively larger amounts of the fund could be diverted to other global environmental and development needs.

The fund is not the usual North-South giveaway. Rather, it represents the transfer of resources from the finance sector—in the form of speculative, nonproductive transactions—to the industrial sector, in the form of intensely productive, wealth-generating, job-creating investments.

The fund also represents a critical investment in our own national security. The global climate envelops us all. What is needed is the kind of thinking that gave rise to the Marshall Plan after World War II. Today, instead of a collection of impoverished and dependent allies in Europe, we have prosperous and robust trading partners. We believe a plan of this magnitude would have a similarly enriching effect on the world's developing economies. It would create millions of jobs. It would raise living standards abroad without compromising ours. It would allow developing countries to grow without regard to atmospheric limits—and without the budgetary burden of imported oil. And in a very short time, the renewable energy industry would emerge as a central, driving engine of growth for the global economy.

Any strategy to change the world's energy diet must directly address the oil-producing nations of the world—especially in the Middle East. Were the countries of the North to phase out their oil consumption without accommodating the historical geopolitical role of the oil-producing nations, the drain on Middle Eastern economies would be disastrous. It would inflame the world's most politically volatile region. It would exacerbate tensions in oil kingdoms where popular resentment against autocratic rulers is already seething. It would strengthen the perception that the United States is waging a war on Islam. Given the high levels of unemployment in Egypt, Iraq, and other Middle Eastern nations, it would elevate economic despair and political desperation to new levels.

The solution to this dilemma lies in the fact that hydrogen will be the central fuel of a new energy economy. The cheapest and most environmentally benign way to make hydrogen is by putting electricity into water and capturing the liberated hydrogen gas.

In this case, an energy modernization plan would involve helping the nations of the Middle East cover their deserts with saltwater pipelines and electricity-generating solar (photovoltaic) panels and wind farms. A structure of vast, hydrogen-producing farms would allow the nations of the Middle East to be hydrogen suppliers to Europe, North Africa, and East Asia. It would allow those countries to use the resources on top of their land (sunlight and wind), instead of the oil deposits below the surface. Although the countries of the Middle East might not realize the same profit margins from hydrogen as they do from oil, they would still retain their geopolitical role as major energy suppliers in the global economy. Moreover, since the installation of wind farms and solar panels is far more labor intensive than the highly automated activity of oil drilling, this kind of initiative would create jobs for many of the underemployed and unemployed people in the region.

One highly improbable fantasy might involve a collaboration between Israel and its Arab neighbors. Israeli scientists and engineers are developing truly cutting-edge solar technologies. Were they to work with their Arab neighbors on a system of solar-powered hydrogen farms, it could begin to create the kind of economic collaboration that might, in time, help create conditions supportive of a real peace process.

The third—and unifying—strategy of the plan—which makes it all work—calls on the parties to Kyoto to subordinate the uneven and inequitable system of international emissions trading to the simple and equitable progressively more stringent Fossil Fuel Efficiency Standard, which would go up by about 5 percent per year.

This mechanism, if incorporated into the Kyoto Protocol, would harmonize and guide the global energy transition in a way that emissions trading cannot.

As noted earlier, the mechanism of emissions trading can work relatively well within nations. Domestic cap-and-trade programs—like the U.S. trading program set up to reduce sulfur dioxide emissions—have been relatively successful because they are easy to monitor and enforce. For that reason, a well-constructed, properly monitored "cap-and-trade" system could work well within individual companies and countries as a supplementary mechanism to help meet the 5 percent annual increase in the Fossil Fuel Efficiency Standard.

Under this mechanism, every country would start at its current baseline to increase its fossil fuel energy efficiency by 5 percent every year until the global 70 percent reduction is attained. That means a country would produce the same amount of goods as the previous year with 5 percent less carbon fuel. Alternatively, it would produce 5 percent more goods with the same carbon fuel use as the previous year.

Since no economy can grow at 5 percent for long, emissions reductions would outpace long-term economic growth.

The fact that every country would begin at its current baseline would eliminate the equity controversies inherent in the "cap-and-trade" system—and would, in tandem with the fund—assure the participation of developing countries.

For the first few years of the efficiency standard, most countries would likely meet their goals by implementing low-cost or even profitable efficiencies—the "low-hanging fruit"—in their current energy systems. After a few years, however, as those efficiencies became more expensive to capture, countries would meet the progressively more stringent standard by drawing more and more energy from noncarbon sources—most of which are 100 percent efficient by a fossil fuel standard.

That, in turn, would create the mass markets and economies of scale for renewables that would bring down their prices and make them competitive with coal and oil.

This approach would be far simpler to negotiate than the current protocol, with its morass of details involving emissions trading, reviews of the adequacy of commitments, and differ-entiated targets. It would also be far easier to monitor and en-force. A nation's compliance would be measured simply by calculating the annual change in the ratio of its carbon fuel use to its gross domestic product. That ratio would have to change by 5 percent a year.

This approach has a precedent in the Montreal Protocol, under which companies phased out ozone-destroying chemi-cals. That protocol was successful because the same compa-nies that made the destructive chemicals were able to produce their substitutes—with no loss of competitive standing within the industry. The energy industry must be reconfig-ured in the same way. Several oil executives have said in pri-vate conversations that they can, in an orderly fashion, decarbonize their energy supplies. It may be true that today British Petroleum is the world's largest vendor of solar sys-tems. Likewise, Shell may have invested $1 billion in a renew-able energy subsidiary. And Ford and Daimler-Chrysler may have entered into a $1 billion joint venture to produce fuel-cell cars. But for these efforts to have more than a marginal impact, the oil and auto industries need the governments of the world to regulate the process so all companies can make the transition in lockstep without losing market share to competitors. The progressive Fossil Fuel Efficiency Standard would, I think, provide that type of regulation.

The plan, then, would be driven by three engines: The subsidy switch would propel the metamorphosis of oil com-panies into clean-energy companies; the competition for the new $300-billion-a-year market in clean energy would power the whole process; and the progressive Fossil Fuel Efficiency Standard would harmonize the transformation of national energy structures, create a predictable regulation for the major energy corporations, and jump renewable energy into a global industry.

A global energy transition requires the governments of the world to regulate some of the largest corporations on the planet. On the record, corporations reflexively resist any move toward new regulation. But history indicates that if the regulations are nondiscriminatory, industry-wide and, most important, predictable—so corporations can depend on them in formulating their strategic plans—business leaders will accept them.

These climate solution strategies present a clear deal to the multinational oil majors: the relinquishing of a measure of corporate autonomy in exchange for a new $300-billion-a-year market.

They present something else as well—the glimpse of an opportunity to begin to democratize the global economy by putting people back in charge of governments and governments in charge of corporations.

The rise of the current corporate state seems, to many people, to be killing off the democratic impulse in the name of economic rationalization. Many people today feel that their human roles have been reduced to agents for the movement of money. Political participation in the United States—especially at the national level—has declined to all-time lows. One powerful reason for this malaise of political alienation lies in the control of the government by large corporate interests. Energized people who organize around obvious and needed reforms—campaign finance reform, environmental protection, worker rights—become exhausted and demoralized as each popular demand for reform is trumped in Washington by lobbyists representing huge corporate interests, who dominate the invisible workings of our national government.

It is not a stretch to say that the solution to the climate crisis contains the potential to begin to reverse our slide into a permanent corporate state and to resuscitate participatory democracy as an operating principle of our civic lives.

Given the reach of the major multinational energy companies in today's globalized economy, it is apparent that the problem requires a mechanism of international governance to achieve this regulation. In the case of the climate crisis, that institution is the U.N. Framework Convention on Climate Change, the compact under which the Kyoto Protocol was drafted. That mechanism of international governance could well provide a pilot model for a much more extensive regime of popular governance of the world's industries.

Even as we are seeing the globalization of the economy, we are also seeing the globalization of communication among activist groups all over the world. In many countries, moreover, governments depend on the expertise and insights of nongovernmental organizations for policy guidance. This is especially true in many developing countries, where governments simply do not have the resources to develop their own expertise in specialized areas. There are also a large number of established NGOs in industrial nations that exert considerable influence on government policies. One can readily imagine a coordinated global movement in which activist organizations around the world mount simultaneous campaigns to pressure their individual governments to impose uniform regulations on multinational industries—and to penalize governments that do not comply.

Those coordinated national-level campaigns could be paralleled by a coordinated international effort. The climate crisis—like many other global problems—requires an international agency to regulate multinational corporations. That agency could provide an access point for real influence by a global community of electronically linked activists. The United Nations has long provided institutional roles for accredited nongovernmental organizations in its decisionmaking. Civil society groups have participated with the U.N. and its member states on a range of issues, including, among others, disarmament, human rights, sustainable development, social policy, and humanitarian affairs.

Initially, one can envision the activist community participating in international decisions in the narrow area of energy technologies. Ultimately, that participation could expand to include popular regulation of many other areas of corporate activities that affect not only our environment but our social and economic well-being as well.

If such a collaboration on climate and carbon emissions were to be broadened and refined, one could imagine people around the world voting on what is the acceptable discrepancy between the salary of a CEO and a janitor. One could imagine people voting to put democratically determined limits on the activities of entire industries. For instance, people might vote to require multinational corporations, setting up manufacturing facilities in developing countries, to pay new workers 150 percent of median income in the host country. One could imagine people voting on a mandatory regime of testing for any new chemical released into the environment. Groups representing investors might vote on a set of corporate reporting standards that are far more transparent and publicly accessible than those that provided cover for the unconscionable deceptions at Enron, Tyco, and WorldCom, among others. In all kinds of areas, people could be given the opportunity to vote on regulations that would limit the most damaging impacts of corporate activities on vulnerable communities, populations, and ecosystems.

This is one promise embedded in the climate crisis—an opportunity to begin to put corporations in the service of people rather than keeping people at the mercy of increasingly huge conglomerates whose activities are determined by the demands of a market that does not account for the social or environmental consequences of its operations.

This is not socialism. Companies would still compete within these democratically determined boundaries—and the quality of product development, customer service, and corporate innovation would continue to determine the winners and losers in the marketplace. In specific industries, these regulations might reduce overall profit margins—but they would be reduced equally for all the companies within an industry.

It seems highly probable, moreover, that those lower profits would be offset by the economic benefits of better working conditions, higher wages, less costly environmental cleanups, far lower expenditures on negative public health impacts, and a massive savings in the amount of money most corporations pay—directly to their lobbyists and indirectly to candidates and their parties—to procure special advantages in Washington.

That is a process that could begin with a program of public involvement in the area of emissions reduction—and lead toward a new kind of democratization of the global economy.

Were the nations of the world to come together around the climate crisis, it could also set the tone for a new set of international relationships. The meltdown of the planet cannot be reversed by unilateral policies or exclusive alliances. The global climate does not respect national boundaries or international coalitions. The climate crisis pits us all against the gathering fury of nature.

But there is a central conundrum embedded in these solutions—how to expand the overall wealth in the global economy without destroying the physical environment on which it depends. Ultimately, the democratization of the economy must be yoked to the twin goals of equity and sustainability. The ominous arrival of climate change aside, the physical planet will not long support global levels of consumption, material use, and pollution that characterize the affluence of the industrial world.

We are cutting down forests at an astonishing rate. Between 1980 and 1995, for instance, at least 2 million square kilometers of forests were destroyed—an area larger than

Mexico. According to a recent study, 90 percent of all large fish have disappeared from the world's oceans in the past half century because of industrial-scale fishing. Each year, we release thousands of new chemicals into the environment—which have already produced growing "dead zones" on ocean floors, contamination of scarce drinking water sources, and an extraordinary increase in the amount of chemicals (and, presumably, cancers, birth defects, and reproductive disorders) in our own bodies over the past several generations.

At the same time that we are decimating our natural world, we are relentlessly exacerbating the ever-widening divide between the world's rich and poor. This is a recipe for certain political chaos.

Overshooting the capacity of the atmosphere to absorb our carbon emissions—and thereby beginning to destabilize the global climate—brings us up against only the first of many limits of the natural world.

A switch to a clean-energy economy will, at best, only buy us a bit more time to prepare for other retaliations of nature that wait just around the corner.

The unprecedented challenge is to raise living standards abroad while simultaneously reducing the rate at which we are depleting the global storehouse of resources and exhausting the capacity of the physical world to accommodate our wastes. Ideally, most of the economic benefits of an energy transition should accrue to the developing world so that, even as the wealth gap becomes progressively narrower, the larger global economy moves toward new levels of sustainability in which raw material is used more sparingly and strategically and in which far more of our waste is eliminated at the source and reused at its endpoint. In the case of the climate crisis, a global transition to clean-energy sources should intrinsically generate more economic growth in the developing world—if only because that would provide more new energy sources to poor countries. The same transition in the industrial world, by contrast, would likely involve a much higher proportion of replacements for currently existing energy sources.

Beyond the technological change involved in a clean-energy revolution, the continuity of a cohesive civilization seems also to require a profound shift in our values. It suggests that much more of our gratification—especially in the wealthier nations—must come from sources other than the acquisition and consumption of an endless stream of products, most of which depend on artificially created demand and many of which are superfluous to our personal happiness. It seems to require, instead, that we look to our intellectual pursuits, our creative activities, our recreational competitions, and our expanding web of relationships for personal fulfillment.

Although this is beginning to happen very gradually—through the shift from an industrial to an information economy—it is happening far too slowly. The transition needs to be accelerated dramatically. The planet will not accommodate anything less. And it is the planet, ultimately, that holds our history hostage against a future that, to remain coherent, must unfold in ways that accommodate nature's limits and thresholds.

These "climate solutions" represent a technological fix to the climate crisis. But given the rapid rate at which many of the systems of the biosphere are deteriorating, it is clear we cannot "fix" our way into a secure future with technological innovations alone.

My own instinct is that changes in values frequently follow changes in technology. The larger hope here is that the very act of addressing the climate crisis in its true proportions would bring home to everyone around the world the realization that we are living on a planet with limits—and that we are now bumping up against those limits.

Ultimately, a worldwide crash program to rewire the world with clean energy would, I believe, yield far more than a fuel switch. It could very easily lead to "closed-loop" industrial processes (which capture industrial waste products rather than releasing them into the environment), "smart-growth" planning (with its emphasis on clustered housing, more open space, and a reliance on public transportation), far more recycling and reuse, the adoption of "environmental accounting" (which includes the true costs of resource depletion and pollution in calculating national GDPs), and, eventually, a whole new ethic of sustainability that would transform our institutions, practices, and dynamics in ways we cannot begin to imagine.

I think the realization that we are all part of a larger—and increasingly vulnerable—community could engender a new sense of common purpose, which would begin with an energy transition and lead, in turn, to a sustainable redesign of the entire human enterprise in a global project that could keep us all very busy for years to come.

This crisis requires the kind of cooperative global response that the founders of the United Nations never imagined. It requires us to abandon the artificial divisions of a stale nationalism that, even against our will, we are outgrowing. Our context is changing faster than our understanding of ourselves.

In the early twenty-first century, the United States is redefining itself as an empire. But every empire is doomed by the limits of its military power. We cannot protect a complex, highly organized society against terror attacks. In the end, all empires contain the seeds of their own destruction—resistance, resentment, and sabotage.

If the United States were to lead the world in a global partnership of this scale, it could lay the foundations for a new era of history. The United States could regain its position of leadership—not through its military power but through its nurture and support of the rest of the world. The payback would take the form of an expansion in trade and commerce as well as a resurrection of moral leadership and international goodwill.

A properly framed global public works program to rewire the world could unite every country around a project that is critical to our common survival. It would not pit countries or alliances against one another. There is no historical precedent for this kind of global cooperation. It has never happened before. The rapid and unpredictable pace of intensifying climate change may never allow it to happen again.

Human institutions and planetary systems operate on very different time schedules. Large planetary systems, with huge amounts of inertia, maintain their equilibrium for centuries, frequently for millennia. But when they do begin to move toward a new state, their momentum may be unstoppable until they settle into a new equilibrium—one that will probably make conditions on the planet far less hospitable to organized civilization.

Human political institutions have much the same type of inertia—albeit on much smaller time scales. By the time we detect the beginnings of really major movements in natural systems, our human institutions must respond immediately.

In the case of the climate crisis, this is happening only partially in other parts of the world—and in the United States barely, if at all. This disjunction may be the undoing of us all. The escalating pace of climate change may well outstrip our capability to respond. We have become so engaged in a moment-to-moment global lifestyle, it seems as though our sense of the future is becoming progressively more truncated.

Today, a significant number of scientists are now saying it is already too late to avoid major disruptions. Many cite extrapolations of current trends. But even the ominous trajectory of those trends pales in the face of a seldom-acknowledged truth about the climate. Nature is capable of immense and unpredictable surprises—and those surprises are becoming progressively more likely than not. The increasing likelihood of major disruptions, of abrupt discontinuities, may render current extrapolations obsolete.

All of which raises the question: Has humanity already passed the point of no return in its slide toward climate chaos?

Several years ago, I was invited to make a short presentation to a small number of senators and congressmen in the office of the late Senator Paul Wellstone of Minnesota. At the end of my presentation, one congressman, John Olver, of Massachusetts, a former chemistry professor, said: "I agree with everything you've just said. Except that I think it's too late."

Although that sentiment—which is echoed by a growing number of observers—is profoundly discouraging, it may not be completely true.

What is true is that it now *appears* very likely that it is too late to avert a cascade of major and destructive impacts of climate change. A great many signals indicate that events are outpacing our ability to contain them. That is the conclusion one reaches from a steady flow of scientific findings and a succession of warming-driven impacts around the world.

But the honest truth is that we really do not know.

We do not know where on the trajectory of disintegration we stand. We cannot identify thresholds of carbon concentrations that could flip the climate into a new regime. We do not know what other feedback mechanisms lie in wait—and when they may kick in.

There is one other unknown that may be even more critical than the mysterious timetables of nature. It has to do with sudden and unpredictable eruptions of sweeping social and political movements.

Given the urgency and magnitude of the escalating pace of climate change, the only hope lies in a rapid and unprecedented mobilization of humanity around this issue. There are a few precious precedents in our recent history. The Berlin Wall fell within a couple of years of the demise of the Soviet Union. The citizens of South Africa overturned that country's long legacy of apartheid in the historical blink of an eye.

This hope represents the most intellectually honest consolation when virtually all the evidence points toward the increasing inevitability of catastrophe: that some spark might ignite a massive uprising of popular will around a unifying movement for social survival and the promise it holds for a more prosperous, more equitable, and more peaceful world. Absent that spark, the prognosis is deeply disheartening. The antidote to the paralysis of despair lies in acknowledging our ignorance. Regardless of the apparent hopelessness of our situation, we really do not know the timing or the nature of the huge surprises embedded in nature and, one hopes, in our own collective behavior. The existential imperative in the face of this profound ignorance is simply to keep trying. No alternative seems morally acceptable.

At this extraordinary cross point in history, we are fast approaching a unique pivot point in our social evolution. Either we will move forward toward a much more cooperative and coordinated global community or we will regress into a progressively more tribalized, combative, and totalitarian existence. And we will watch the unbounded promise of the future—

which has been our birthright since the beginning of civilization—as it disintegrates in a cascade of climatic disruptions.

There is really no choice. One way or the other, this world we inhabit will not long continue on its historical trajectory. Like it or not, we are facing a massive and inevitable discontinuity.

We will either retreat into ourselves and scramble to defend our private security in an increasingly threatening environment, or we will move forward into a much more coherent and prosperous and peaceful future.

This perspective may well be overly visionary. But the alternative—given the escalating instability of the climate system, the deterioration of other natural planetary systems, and the increasing desperation of global economic inequity—is truly horrible to contemplate.

The ultimate hope is that—especially given the centrality of energy to our modern lives—a meaningful solution to the climate crisis could potentially be the beginning of a much larger transformation of our social and economic dynamics.

Our modern history has been marked by a dichotomy between the totalitarianism of command-and-control economies and the opulence and brutality of unregulated markets and runaway globalization.

It is just possible that the act of rewiring the planet could begin to point us toward that optimal calibration of competition and cooperation that would maximize our energy and creativity and productivity while, at the same time, substantially extending the baseline conditions for peace—peace among people, and peace between people and nature.

Notes

Chapter 1: Not Just Another Issue

1 **We've known for some time** Margaret Beckett, British secretary of state for environment, "Record Warm Start to 2002," http://www.BBCNews.com, April 26, 2002.

4 **In 2001, researchers at the Hadley Center** "Acceleration of Global Warming due to Carbon-Cycle Feedbacks in a Coupled Climate Model," Peter M. Cox et al., *Nature* 408, November 9, 2000; "Expert Warns World Warming Faster Than Expected," Reuters News Service, May 13, 2002.

4 **Three years ago** "Energy Implications of Future Stabilization of Atmospheric CO_2 Content," Martin I. Hoffert et al., *Nature* 395, October 29, 1998.

4 **In 2002, a follow-up study** "Advanced Technology Paths to Global Climate Stability: Energy for a Greenhouse Planet," Martin I. Hoffert et al., *Science* 298, November 1, 2002.

5 **In 1995, the issue gained** J. T. Houghton et al., *Climate Change—1995: The Science of Climate Change: Contribution of Working Group I to the Second Assessment Report of the Intergovernmental Panel on Climate Change* (Cambridge: Cambridge University Press, 1996).

5 **That January** J. T. Houghton et al., *Climate Change 2001: The Scientific Basis, Third Assessment Report of the Intergovernmental Panel on Climate Change* (Cambridge: Cambridge University Press, 2001).

5 **The most comprehensive study** "Scientists Issue Dire Prediction on Warming, Faster Climate Shift Portends Global Calamity This Century," *Washington Post*, January 23, 2001.

6 **A second working group** "U.N. Report Forecasts Crises Brought On by Global Warming, Poor Countries Would Bear Brunt of Climate Consequences," *Washington Post*, February 20, 2001; Intergovernmental Panel on Climate Change, *Climate Change 2001: Impact, Adaptation, and Vulnerability* (Cambridge: Cambridge University Press, 2001).

6 **Two years later** "Climate Change 'Cost $60b' in 2003," http://www.CNN.com, December 12, 2003.

6 **The scientific consensus** "Scientists Issue Dire Prediction on Warming, Faster Climate Shift Portends Global Calamity This Century," *Washington Post*, January 23, 2001.

7 **The entire ecosystem** "North Sea Faces Collapse of Its Ecosystem, Fish Stocks and Sea Bird Numbers Plummet as

Soaring Water Temperatures Kill Off Vital Plankton," *Independent* (UK), October 19, 2003.

7 **For the first time** "World Facing Fourth Consecutive Grain Harvest Shortfall," Lester R. Brown, Earth Policy Institute, September 2003.

7 **The German government declared** "Melting Ice 'Will Swamp Capitals,'" *Independent* (UK), December 7, 2003.

7 **The most highly publicized** "Maximum and Minimum Temperature Trends for the Globe," David R. Easterling et al., *Science* 277, July 18, 1998.

8 **In August 2003** "European Heatwave Caused 35,000 Deaths," *NewScientist*, October 10, 2003.

8 **The following month** "Arctic's Biggest Ice Shelf, a Sentinel of Climate Change, Cracks Apart," *Los Angeles Times*, September 23, 2003.

9 **The same month** "Oceans' Acidity Worries Experts, Report: Carbon Dioxide on Rise, Marine Life at Risk," *Atlanta Journal-Constitution*, September 25, 2003.

9 **By the fall of 2003** "Scientists See Antarctic Vortex as Drought Maker," Reuters News Service, September 23, 2003.

9 **Pension fund managers** "Pension Funds Plan to Press Global Warming as an Issue," *New York Times*, November 22, 2003.

10 **The meeting was** Ibid.

10 **In May 2003, Swiss Reinsurance** "Insurers Weigh Moves to Cut Liability for Global Warming, Directors, Officers Could Face the Denial of Coverage After Rules Are Implemented," *Wall Street Journal*, May 7, 2003.

10 **In late 2003, the oil giant** "Global Warming Gas Seen Increasing Dramatically," Reuters News Service, November 19, 2003.

11 **Unlike the world's** Ibid.

11 **The final—and perhaps** "Kremlin Aide Sows Confusion as Rejects Kyoto Again," Reuters News Service, December 5, 2003.

11 **With the United States already** "Russia: Show Me the Money," *Asia Times*, December 9, 2003.

Snapshots of the Warming No. 1

19 **Scientists had documented** "200-Metre Thick Antarctic Ice Shelf Melts," *Guardian* (UK), March 20, 2002.

20 **Because ice shelves float** "Study of Antarctic Points to Rising Sea Levels," *New York Times*, March 7, 2003; "Greenland's Warming Ice Flows Faster," http://www.BBCNews .com, June 7, 2002; "Study of Antarctic Points to Rising Sea Levels," *New York Times*, March 7, 2003.

21 **A National Aeronautics** "Arctic Ice Melting Much Faster Than Thought, NASA Study Shows About 9 Percent Is Disappearing Every 10 Years," *Toronto Globe and Mail*, November 28, 2002.

21 **For example, the Arctic ice shelf** "Arctic's Biggest Ice Shelf, a Sentinel of Climate Change, Cracks Apart," *Los Angeles Times*, September 23, 2003; "Largest Arctic Ice Shelf Breaks Up, Scientists Say," Reuters News Service, September 22, 2003; "Huge Ice Shelf Is Reported to Break Up in Canada, *New York Times*, September 23, 2003.

21 **In the Andes** "As Andean Glaciers Shrink, Water Worries Grow," *New York Times*, November 24, 2002.

21 **Glacier National Park** "Glacial Retreat Faster Than Previously Thought," *Boston Globe*, May 28, 1998; "Glacier Park on Thin Ice," *Los Angeles Times*, November 18, 2002.

22 **In Kenya** "Kilimanjaro's Ice 'Archive,'" http://www.BBCNews.com, October 18, 2002.

22 **In India** "Heat on the Mountain," *Indian Sunday Express*, September 15, 2002.

Chapter 2: The Sum of All Clues

23 **Prehistoric and early historic** "What Drives Societal Collapse," Harvey Weiss and Raymond S. Bradley, *Science*, January 26, 2001.

25 **That year, a team of researchers** B. D. Santer et al., "A Search for Human Influences on the Thermal Structure of the Atmosphere," *Nature* 382, July 4, 1996.

25 **A second "smoking gun"** Thomas Karl et al., "Trends in U.S. Climate During the Twentieth Century," *Consequences* 1 (1) (Spring 1995); see also, Thomas R. Karl, Neville Nicholls, and Jonathan Gregory, "The Coming Climate," *Scientific American*, May 1997.

26 **A third contribution** David J. Thompson, "The Seasons, Global Temperature, and Precession," *Science* 268, April 7, 1995.

26 **In 1997, a research team** David Easterling et al., "Temperature Range Narrows Between Daytime Highs and Nighttime Lows," *Science*, July 18, 1997.

27 **In 1999, a team** Phil Ball, "Reading the Signs," *Nature* 399, June 10, 1999.

27 **A year later, Thomas Crowley** Thomas J. Crowley, "Causes of Climate Change over the Past 1,000 Years," *Science* 289, July 14, 2000.

28 **The Crowley findings** "Northern Hemisphere Temperatures During the Past Millennium: Inferences, Uncertainties, and Limitations," Michael E. Mann, Raymond S. Bradley, and Malcolm K. Hughes, *Geophysical Research Letters* 26 (6), March 15, 1999.

28 **They found that** "Researchers Confirm Millennium Finding, 1998 Was Warmest Year of Millennium, Climate," press release of the American Geophysical Union, March 3, 1999.

28 **That year, scientists studying** "Increase in Greenhouse Gases Seen from Space," Reuters News Service, March 14, 2001; John E. Harries et al., "Increases in Greenhouse Forcing Inferred from the Outgoing Longwave Radiation Spectra of the Earth in 1970 and 1997," *Nature* 410, March 15, 2001.

29 **In early 2000** Thomas R. Karl, Richard W. Knight and Bruce Baker, "The Record-Breaking Global Temperatures of 1997 and 1998," *Geophysical Research Letters* 27 (5), March 1, 2000.

29 **Said Jonathan Overpeck** Ibid.

29 **That heating** "Ocean Study Points Finger at Mankind, Older Computer Models Did Not Factor In the Ocean," http://www.BBCNews.com, April 11, 2001.

29 **That year, two teams** Sydney Levitus et al., "Anthropogenic Warming of Earth's Climate System," *Science* 292, April 13, 2001.

30 **The Scripps team** Tim P. Barnett, David W. Pierce, and Reiner Schnur, "Detection of Anthropogenic Climate Change in the World's Oceans," *Science* 292, April 13, 2001.

31 **Modern climate change is dominated** Thomas R. Karl and Kevin E. Trenberth, "Modern Global Climate Change," *Science* 302, December 5, 2003.

32 **Senior scientist Tom M.L. Wigley** letter (in author's possession) from Tom M.L. Wigley to Senators Tom Daschle (D–ND) and William Frist (R–TN), July 2003.

Snapshots of the Warming No. 2

33 **In early 2003, two major studies** "Minute Shift in Temperature Has Had a Major Effect on Earth, Studies Show," *Los Angeles Times*, January 2, 2003.

33 **One study showed** Terry L. Root et al., "Fingerprints of Global Warming on Wild Animals and Plants," *Nature* 421, January 2, 2003.

34 **A second study** Camille Parmesan and Gary Yohe, "A Globally Coherent Fingerprint of Climate Change Impacts Across Natural Systems," *Nature* 421, January 2, 2003; "Global Warming Found to Displace Species," *New York Times*, January 2, 2003.

34 **The report came** Gian-Reto Walther et al., "Ecological Responses to Recent Climate Change," *Nature* 416, March 28, 2002.

35 **There is now ample evidence** "Effects of Climate Warming Already in Evidence," Environmental News Service, March 29, 2002.

35 **In Britain** "Seasons 'Becoming Muddled,'" http://www.BBCNews.com, September 7, 2002.

35 **The change in the timing** Author's conversation with Bill McKibben, 2000.

36 **On landmasses around the world** "Global Warming Threatens One-Third of All Habitat," Environmental News Service, August 30, 2000.

36 **These forecasts** "Warming May Threaten 37 Percent of Species by 2050," *Washington Post*, January 8, 2004; Chris D. Thomas et al., "Extinction Risk from Climate Change," *Nature* 427, January 8, 2004.

Chapter 3: Criminals Against Humanity

37 **With all of the hysteria** Senator James Inhofe (R–OK), July 28, 2003, speech on Senate floor (in author's possession).

38 **By contrast** "Businesses Set Voluntary Plan to Cut Greenhouse-Gas Output," *Wall Street Journal*, February 13, 2003.

39 **More damning, perhaps** *Preliminary Observations on the Administration's February 2002 Climate Initiative*, U.S. General Accounting Office, October 1, 2003.

40 **Shortly before Bush's** *Bureau of National Affairs*, January 24, 2001.

41 **The official line** Frank Luntz memo to the Republican Party, 2002, italics in the original (in author's possession).

42 **In a truly Orwellian** *Harper's Magazine*, September 2003.

43 **For starters** "Many Made the Move from the Industry to the Administration," *New York Times*, April 20, 2002.

44 **One coal executive alone** "A Coal-Fired Crusade Helped Bring Bush a Crucial Victory," *Wall Street Journal*, June 13, 2001.

44 **Overall, coal interests** Ibid.; see also, "Public Information and Policy Research: Total Worldwide Public Information and Policy Research Contributions in 2001 Were $5.7 million," ExxonMobil Web site, 2001; see also, "Exxon Backs Groups That Question Global Warming," *New York Times*, May 28, 2003.

44 **Five months after** "A Coal-Fired Crusade Helped Bring Bush a Crucial Victory," *Wall Street Journal*, June 13, 2001.

44 **That "payback" came** "Bush Reverses Vow to Curb Gas Tied to Global Warming," *New York Times*, March 14, 2001.

45 **One person who** "Coal Scores with Wager on Bush," *Washington Post*, March 25, 2001.

45 **Shortly after taking office** "Bush National Energy Policy Expands Nuclear, Oil Drilling," Environmental News Service, May 2, 2001; *New York Times*, May 1, 2001.

45 **Conservation may be a sign** "Cheney Promotes Increasing Supply as Energy Policy," *New York Times*, May 1, 2001.

45 **The work of Cheney's** "Bush Proposes Energy Plan Tapping Oil and Gas Reserves," *New York Times*, May 17, 2001.

46 **In 1998, O'Neill told** "Heatbeat," *Grist Magazine*, March 14, 2001.

46 **Peabody's management** "Top G.O.P. Donors in Energy Industry Met Cheney Panel," *New York Times*, March 1, 2002.

47 **Overnight its stock jumped** Author's private conversation with an individual familiar with the Peabody IPO offering.

47 **The IPO was handled** "Top G.O.P. Donors in Energy Industry Met Cheney Panel," *New York Times*, March 1, 2002.

47 **On February 6, 2001** Randy Randol, ExxonMobil memo (in author's possession).

48 **In their place** "Dissent in the Maelstrom," *Scientific American*, November 2001.

48 **The satellite argument** F. J. Wentz and M. Schabel, "Effects of Orbital Decay on Satellite-Derived Lower-Tropospheric Temperature Trends," *Nature* 394, August 13, 1998.

48 **Lindzen, for his part** Author's conversation with Richard Lindzen in 1997.

49 **One team of NASA researchers** "Dissent in the Maelstrom," *Scientific American*, November 2001.

49 **ExxonMobil officials also** "Bush Appointed Oil Giant's Candidate," *Telegraph of London*, May 16, 2002.

50 **His position was clear** "U.S. Dashes Hopes for Climate Deal," *Guardian* (UK), May 14, 2002.

50 **In an effort to improve** "ExxonMobil outlines 100-billion-dollar investment plans," Agence France Presse, November 6, 2002.

50 **Two months later** "Hydrogen Powers Bush State of the Union Address," Environmental News Service, January 28, 2003.

51 **ExxonMobil's new public relations** "Exxon Backs Groups That Question Global Warming," *New York Times*, May 28, 2003.

51 **During the 1990s** Ross Gelbspan, *The Heat Is On: The Climate Crisis, the Cover-Up, the Prescription* (Cambridge: Perseus Books, 1998).

51 **Throughout the 1990s** Ibid.

52 **What is especially telling** Tom M.L. Wigley, "A Scientific Critique of the Greenhouse Skeptics," appendix to Gelbspan, *The Heat Is On*.

53 **As for Singer** "Hearing on Scientific Integrity and the Public Trust: The Science Behind Federal Policies and Mandates, Case Study 1," 104th Cong., House Subcommittee on Energy and Environment of the Committee on Science, September 20, 1995.

53 **Singer's recklessness** *Washington Post*, Letters to the Editor, February 12, 2001.

53 **In fact, Singer received** A pdf version of the page from the ExxonMobil Web site (in author's possession); Ross Gelbspan, "Global Warmers," *Nation*, March 20, 2001.

54 **The paper was coauthored** Harvard-Smithsonian Center for Astrophysics, press release: "Twentieth Century Climate Not So Hot," March 31, 2003.

55 **The recent report** Copy of ExxonMobil Web site page, listing funding of the Harvard-Smithsonian Center for

Astrophysics (in author's possession); Harvard-Smithsonian Center for Astrophysics press release: "Twentieth-Century Climate Not So Hot," March 31, 2003.

55 **Nevertheless, the study** "The Science of Climate Change," statement by Senator James M. Inhofe (R–OK), chairman, Committee on Environment and Public Works, July 28, 2003.

55 **It is probably not a coincidence** Web site of the Center for Responsive Politics, http://www.opensecrets.org.

55 **In short order** "Leading Climate Scientists Reaffirm View That Late Twentieth Century Warming Was Unusual," *Eos*, American Geophysical Union, July 8, 2003.

56 **Three editors** Jeff Nesmith, "Two Editors Resign in Protest over Flaws in Paper by Skeptics," Cox News Service, July 29, 2003; "Global Warming Skeptics Are Facing Storm Clouds," *Wall Street Journal*, July 31, 2003. (A third editor resigned following publication of the Cox News Service story.)

56 **The attack on the document** Letter from Center for Regulatory Effectiveness to White House Office of Science and Technology Policy and the Office of Management and Budget, February 11, 2002 (in author's possession).

57 **Not surprisingly, the head** Web site of the Center for Responsive Politics, http://www.opensecrets.org/Lobbyists/lobbyist.asp?ID=23780&year=1999.

57 **The critique** Gelbspan, *The Heat Is On*.

57 **When that avenue** Competitive Enterprise Institute press release, "Group Sues to Enforce Sound Science Law," August 13, 2003.

57 **In August 2003** Press release from Attorney General Steven Rowe of Maine, August 13, 2003.

58 **Nor is it a coincidence** "Public Information and Policy Research: Total Worldwide Public Information and Policy Research Contributions in 2001 Were $5.7 million," from ExxonMobil Web site, 2001; "Exxon Backs Groups That Question Global Warming," *New York Times*, May 28, 2003.

60 **As of September 2003** Web site of U.N. Framework Convention on Climate Change, http://unfccc.int.

Snapshots of the Warming No. 3

63 **In 2002, scientists** "Warmer Water Changing Portugal Fish Species," http://www.Planetark.org, April 2, 2002.

64 **One consequence of** "Ocean Warming Impacts Sea Life Faster Than Expected, Peer-Reviewed Findings Show Effects on Marine Environment Are Starting Earlier, Reaching Farther Than Previously Believed," *World Wildlife Fund* and *Marine Conservation Biology Institute*, June 8, 1999.

64 **Warmer temperatures are** Ibid.

64 **The shrinkage of** "Coral Reefs Are Shrinking Fast—UN Report," Reuters News Service, September 12, 2001.

65 **Although the damage to coral reefs** Ibid.

65 **The ocean warming** "Emerging Marine Diseases—Climate Links and Anthropogenic Factors," C. D. Harvell et al., *Science*, September 22, 1999.

65 **In a startling report** "North Sea Faces Collapse of Its Ecosystem, Fish Stocks and Sea Bird Numbers Plummet as Soaring Water Temperatures Kill Off Vital Plankton," *Independent* (UK), October 19, 2003.

Chapter 4: Bad Press

67 **[T]he press's adherence to balance** Maxwell T. Boykoff and Jules M. Boykoff, "Balance as Bias: Global Warming and the U.S. Prestige Press," preprint version of paper scheduled for publication in *Global Environmental Change* 14 (1) (June 2004).

67 **In 1997, Bert Bolin** Press release of Intergovernmental Panel on Climate Change, Geneva, Switzerland, June 26, 1997, "IPCC Chair Denies Attack on VP Gore, Environmentalists."

68 **Or, as James McCarthy** Gelbspan, *The Heat Is On*.

70 **That contrast is apparent** Anja Kollmuss, "Media Framing of Climate Change: A Cross-Country Analysis of Newspaper Coverage in the United States, the United Kingdom and Germany," MA thesis, Tufts University, 2000.

71 **In June 2003, the European Union** "Europe Adopts Climate Emissions Trading Law," Environmental News Service, July 23, 2003; "EU Strikes Agreement on Emissions Trading," Associated Press, December 10, 2002.

71 **Nor have American journalists** "Wind Turbines Are Sprouting off Europe's Shores," *New York Times*, December 8, 2002.

72 **In fact, many developing** Kilaparti Ramakrishna et al., *Developing Country Initiatives: Action Versus Words: Implementation of the UNFCCC by Select Developing Countries* (Woods Hole, MA: Woods Hole Research Center, 2003).

72 **Even China** "China Said to Sharply Reduce Carbon Emissions," *New York Times*, June 15, 2001.

73 **As James Baker** "Consensus Emerges Earth Is Warming—Now What?" *Washington Post*, November 12, 1997.

76 **Following up on** David R. Easterling et al., "Climate Extremes: Observations, Modeling, and Impacts," *Science*, September 22, 2000.

77 **Those findings were underscored** "Reaping the Whirlwind, Extreme Weather Prompts Unprecedented Global Warming Alert," *Independent* (UK), July 3, 2003; press release from World Meteorological Organization, July 2, 2003.

77 **The destructive power** report by Robert Krulwich, "Elephants in the Sky," *ABC World News Tonight*, September 3, 2003.

78 **At the beginning of the year** "Worst Flooding in at Least 273 Years Paralyzes Parts of Britain," Reuters News Service, November 6, 2000; "Snow Paralyzes Northeast Asia; Mongolia Hit by Deep Freeze," http://www.Weather.com, January 9, 2001; Floridians Slow to Heed Drought Warnings," *Boston Globe*, February 17, 2001; "Firefighters Battle Florida Wildfire," Associated Press, February 20, 2001; "Pakistan Heat Wave Kills 36," http://www.Weather.com, May 7, 2001; "Mayor: Storm City's Biggest Disaster Ever," *Houston Chronicle*, June 14, 2001; "Tropical Storm Allison the Costliest in U.S. History," United Press International, October 30, 2001; "Drought Creates Food Crisis in Central America," *New York Times*, August 28, 2001; "Malnourished to Get Help in Guatemala," *New York Times*, March 20, 2002; "Iran Drought Turns Lakes to Scorched Earth," Reuters News Service, August 1, 2001; "Iran Flood Death Toll Soars, Scores Missing," Reuters News Service, August 13, 2001; "October Tornadoes Set U.S. Record," http://www.Weather.com, October 25, 2001; "Rescuers Fear Toll of 1,000 in Algeria," *Boston Globe*,

November 15, 2001; "Record-Breaking Warmth Stretches Far North," http://www.Weather.com, January 20, 2002; "Record Temps Make for Warm Winter," http://www.Weather.com, January 29, 2002.

79 **In the following year** "Heat Wave in India Kills 1,000," Associated Press, May 22, 2002; "Death Toll in S. Russia Flood Climbs," Associated Press, June 29, 2002; "Four Killed as Storms Batter France and Germany," http://www.Planetark.org, June 10, 2002; "Russian Flash Floods Wreak Havoc," http://www.BBCNews.com, August 9, 2002; "Crews Maxed Out as Fires Spread," http://www.MSNBC.com, June 21, 2002; "Summer of Extremes Baffles Specialists," *Boston Globe*, August 17, 2002; "Drought Spreads with June Heat," http://www.Weather.com, July 22, 2002; "India Grid Collapse Causes Blackout," Associated Press, July 31, 2002; "Malaria-Carrying Mosquitoes Found," Associated Press, September 29, 2002; "West Nile Virus Spreads Across United States," http://www.Planetark.org, July 29, 2002; author's conversations with Dr. Paul Epstein, Center for Health and the Global Environment, Harvard Medical School, and Paul Epstein, "Is Global Warming Harmful to Health?" *Scientific American*, August 2000; "Millions Stranded, 445 Dead in Asia Floods," http://www.Weather.com, July 29, 2002.

79 **In the spring of 2003** "Monsoon Showers Ease India's Blistering Heat Wave," http://www.Reuters.com, June 6, 2003; "U.S. Sets Tornado Record," http://www.CNN.com, May 14, 2003; "From the Alps to Arid Southern Forests, Europe Bakes in Record Heat; London Breaks 100

Degrees," Associated Press, August 10, 2003; "Europe's Heat Wave Raises Global Warming Concerns," Environmental News Service, August 1, 2003; "Europe Blisters Under Heat Wave," http://www.MSNBC.com, July 16, 2003; "Portugal Declares Fire Calamity," http://www.BBCNews.com, August 4, 2003; "Portugal Fire Cost Tops $1 Billion," Reuters News Service, August 9, 2003; "French Heat Deaths up to 3,000," http://www.BBCNews.com, August 14, 2003.

80 **The editor agreed** Author's private conversation with a news network executive, October 1999.

81 **In their paper** Maxwell T. Boykoff and Jules M. Boykoff, "Balance as Bias: Global Warming and the U.S. Prestige Press," paper presented at the 2002 Berlin Conference on the Human Dimensions of Global Environmental Change.

Snapshots of the Warming No. 4

87 **Separate and apart** Kevin E. Trenberth, "The Extreme Weather Events of 1997 and 1998," *Consequences* 5 (1) (1999): 3–15.

88 **Two years later** "Coral Reef Exposes Worst El Niños Ever Are Now," Reuters News Service, January 28, 2001; Alexander W. Tudhope et al., "Variability in the El Niño Southern Oscillation Through a Glacial-Interglacial Cycle," *Science*, January 26, 2001.

89 **In fact, the El Niño of 1997–1998** Trenberth, "The Extreme Weather Events of 1997 and 1998."

89 **Even as they are heating** "Oceans' Acidity Worries Experts," *Atlanta Journal-Constitution*, September 25, 2003;

"Alarm over Acidifying Oceans," *New Scientist,* September 25, 2003.

90 **Their findings indicated** Ken Caldeira and Michael E. Wickett, "Anthropogenic Carbon and Ocean pH," *Nature* 425, September 25, 2003.

90 **Projected climate changes** Intergovernmental Panel on Climate Change, *Climate Change 2001: Impact, Adaptation and Vulnerability.*

90 **According to scientists** Author's conversation with Dr. Paul Mayewski, University of New Hampshire, 1999.

91 **The natural warming** Author's conversation with Barrett Rock, director, Complex Systems Research Center, University of New Hampshire, 2003.

91 **The very recent freshening signal** Robert B. Gagosian, president and director of Woods Hole Oceanographic Institution, writing on the institution's Web site: http://www.whoi.edu, 2002.

91 **At the end of 2003** "Saltier Atlantic May Help Decipher Global Warming," *Boston Globe,* December 18, 2003; Ruth Curry, Bob Dickson, and Igor Yashayaev, "A Change in the Freshwater Balance of the Atlantic Ocean over the Past Four Decades," *Nature* 426, December 18, 2003.

92 **Added Ransom Myers** "Atlantic's Salt Balance Poses Threat, Study Says," *Toronto Globe and Mail,* December 18, 2003.

92 **As Terry Joyce** Joyce quoted in article by Robert B. Gagosian, president and director of Woods Hole Oceanographic Institution: http://www.whoi.edu, 2002.

92 **A subsequent study** "Global Warming Will Plunge Britain into New Ice Age 'Within Decades,'" *Independent* (UK), January 25, 2004.

Chapter 5: Three Fronts of the Climate War
93 **We are all adrift** Raul Estrada Oyuela, Kyoto, Japan, December 1997.

94 **The letter made it clear** "EU Says Climate Strategy 'Integral' to U.S. Relations," Reuters News Service, March 23, 2001.

95 **The president's response** "U.S. Abandons Kyoto Climate Pact, a Blow to Europe," Reuters News Service, March 29, 2001.

95 **The European diplomats were** "Bush Emissions Policies Upsetting Allies Abroad," *Boston Globe*, March 29, 2001.

95 **Swedish prime minister Goeran Persson** "Bush Faces Environmental Opposition and Protesters in Sweden," Associated Press, June 14, 2001.

96 **Ironically, just weeks earlier** "U.S. Aims to Pull Out of Warming Treaty, 'No Interest' in Implementing Kyoto Pact, Whitman," *Washington Post*, March 28, 2001.

96 **The demonstrators' anger** "Allies Tell Bush They'll Act Alone on Climate Accord," *New York Times*, July 22, 2001.

96 **Added French president Jacques Chirac** Ibid.

97 **Despite U.S. intransigence** "EU Ratifies Global Warming Pact," http://www.BBCNews.com, May 31, 2002.

97 **The ratification by the European Union** "EU Agrees Tough Stance Before Climate Change Talks," Reuters News Service, November 8, 2000.

98 **As Sen. James Jeffords** "White House Undermining Environment Summit-Senator," Reuters News Service, July 24, 2002.

99 **After a week of deadlocks** "U.S. Reaches Energy Deal at Summit, No Timetable for Alternatives to Fossil Fuels," *Boston Globe*, September 3, 2002.

99 **The U.S. decision on Kyoto** Jessica T. Mathews, "Estranged Partners," *Foreign Policy* (November–December 2001).

101 **In the case of climate change** "U.S., Australia Climate Plan Cuts No Emissions," Environmental News Service, July 12, 2002.

101 **Sunita Narain** Sunita Narain, "How Not to Lose All," *Down to Earth* (October 2003).

102 **Kyoto is** "U.S. Blasts Kyoto Pact, UN Hopes Russia Will Sign," Reuters News Service, December 1, 2003.

102 **The Bush administration's desire** "Russia Pulls Away from Kyoto Pact," http://www.BBCNews.com, December 2, 2003.

103 **That plan was drafted** "Businesses Set Voluntary Plan to Cut Greenhouse-Gas Output," *Wall Street Journal*, February 13, 2003.

103 **In short order, the Washington-based** Ibid.

104 **Europe's anger at** "U.S. Climate Policy Bigger Threat to World Than Terrorism," *Independent* (UK), January 9, 2003.

105 **In late 2002, the Pew Center** "On Global Warming, States Act Locally," *Washington Post*, November 11, 2002.

106 **In July 2002, the attorneys general** "State Officials Want Bush to Act on Global Warming," Reuters News Service, July 18, 2002.

107 **Several months later** Statement by U.S. Conference of Mayors: "U.S. Mayors Disagree with Parts of Bush Agenda," Environmental News Service, June 30, 2002.

107 **When the Bush administration** "A Pollutant by Any Other Name," editorial, *New York Times*, February 22, 2003.

107 **That suit, which was still pending** "States Sue the Federal Government to Control Greenhouse Emissions," *Wall Street Journal*, October 23, 2003.

108 **On the West Coast** "Statement of the Governors of California, Oregon and Washington on Regional Action to Address Global Warming," September 22, 2003 (in author's possession).

109 **Another oil giant** "Shell Sees Growing Role for Natgas, Renewables" Reuters News Service, February 14, 2001.

109 **Shell followed that** "Shell Extends Clean Energy Push," Reuters News Service," June 15, 2001.

110 **Later that year, ExxonMobil** "Global Warming Gas Seen Increasing Dramatically," Reuters News Service, November 19, 2003.

110 **As ExxonMobil's president** Interview with Lee Raymond in *Chief Executive Magazine*, October 2002.

111 **The split within** "Bush Gets Heat on Global Warming, Entergy Chief Urges Reducing Emissions," *New Orleans Times-Picayune*, September 11, 2003.

112 **Emissions reductions are** "Insurers Weigh Moves to Cut Liability for Global Warming," *Wall Street Journal*, May 7, 2003.

114 **Said California treasurer** "Pension Funds Plan to Press Global Warming as an Issue," *New York Times*, November 22, 2003.

114 **In June 2002** "State Farm Curbs Home Policies," Reuters News Service, June 21, 2002.

114 **Gene Lecomte was** Gene Lecomte, remarks made at "Reporting on Nature's Deadline," conference for news editors at Tufts University, January 2003.

115 **What many bondholders may not realize** "U.S. Hit with 42 Billion Dollar Loss Events Since 1980," Reuters News Service, August 6, 1999.

115 **Noting that about half** Lecomte, "Reporting on Nature's Deadline"; also, Lecomte conversation with author, December 2002 and January 2003.

117 **As BP chief John Browne** "More Light, Less Heat: If Countries and Companies Are Willing to Co-Operate, Global Warming Can Be Brought Under Control," opinion column by John Browne, *Financial Times,* April 2, 2002.

Snapshots of the Warming No. 5

119 **In 2002, a team of researchers** C. Drew Harvell et al., "Climate Warming and Disease Risks for Terrestrial and Marine Biota," *Science*, June 20, 2002.

120 **As the *Boston Globe* reported** "Disease Threat Cited in Global Warming, Report Predicts Virulence and Range Will Grow," *Boston Globe*, June 21, 2002.

120 **The researchers reported** Harvell et al., "Climate Warming and Disease Risks for Terrestrial and Marine Biota."

120 **What is most surprising** "Study: Warmer Climate, Sicker Earth, Cross-Species Observations See Climate/Disease Connection," http://www.MSNBC.Com, June 20, 2002.

121 **This isn't just a question** Ibid.

121 **In Canada, an explosion** "Canadian Pine Beetle Epidemic Now 'Catastrophic,'" http://www.Planetark.org, November 25, 2002.

122 **About 160,000 people** "160,000 Said Dying Yearly from Global Warming," Reuters News Service, October 1, 2003.

122 **There is growing evidence** "Climate Change 'Will Harm Health,'" http://www.BBCNews.com, December 11, 2003.

122 **Mosquitoes, which historically** Paul R. Epstein, "Is Global Warming Harmful to Health?" *Scientific American*, August 2000; author's conversations with Dr. Paul R. Epstein, assistant director, Center for Health and the Global Environment, Harvard Medical School.

122 **In Africa alone, malaria is** "World Watches As Malaria Death Toll Rises," Environmental News Service, April 28, 2003.

123 **According to an article** Epstein, "Is Global Warming Harmful to Health?"

123 **Mosquito-borne disorders** Ibid., pp. 121–122.

123 **Epstein, who is assistant director** Ibid.

125 **It will also, scientists report** "U.K. Faces Summers of Malaria," BBC News, January 22, 2002.

125 **Dengue or "breakbone" fever** Epstein, "Is Global Warming Harmful to Health?"

125 **Another insect that flourishes** "Warmer Sweden Linked with Tick-Born Encephalitis," Reuters News Service, July 9, 2001.

125 **The reasons** "Wet Spring Increases Lyme Risk," *Boston Globe*, July 4, 2003.

126 **The changes in the climate** Fakhri Bazzaz, Peter Wayne, and Susannah Foster, *Annals of Allergy, Asthma and Immunology* 88 (March 2002).

126 **Two years ago** "Global Warming Seen as Doubling Heat Deaths by 2020," Reuters News Service, November 22, 2000.

Chapter 6: Compromised Activists

127 **On climate change** "Blair Speaks Out Against U.S. Refusal to Ratify Kyoto," *Independent* (UK), September 2, 2002.

131 **Ironically, though most** "Climate Collapse: The Pentagon's Weather Nightmare," *Fortune Magazine,* January 26, 2004.

132 **These groups are running around** Author's conversation with Dianne Dumanoski, 2003.

133 **Amory Lovins, director** Amory Lovins et al., *Hypercars: Materials, Manufacturers, and Policy Implications* (Snowmass, CO: Rocky Mountain Institute, 2004).

135 **For example, Environmental Defense** http://www.environmentaldefense.org.

135 **Virtually all the same** http://www.cool-campus.org.

137 **Normally, dissident shareholder** Private conversation between the author and ExxonMobil officials, June 2002.

144 **As Alden Meyer, an energy policy specialist** Author's conversation with Alden Meyer, 1997; see also, Gelbspan, *The Heat Is On.*

144 **an indifference that the British medical journal** "Climate Change—the New Bioterrorism," *The Lancet* 358 (9294), November 17, 2001.

145 **Rising sea levels** "1,000 Flee as Sea Begins to Swallow Up Pacific Islands," *Independent* (UK), November 29, 2000.

145 **The linkage between** Address by Ambassador Hurst to the "Small Island Countries' Dialogue on Water and Climate," World Water Forum, Kyoto, Japan, March 16–23, 2003.

Snapshots of the Warming No. 6

147 **Higher temperatures are thwarting** Lester R. Brown, "World Facing Fourth Consecutive Grain Harvest Shortfall," Earth Policy Institute report, September 2003.

148 **The prognosis for the future** L. O. Mearns et al., "Climate Scenarios for the Southeastern U.S. Based on GCM and Regional Model Simulations," *Climatic Change* 60 (September 2003).

149 **As the world's food supplies** "Linking Environment and Health: Malnutrition," World Resources Institute, 1998–1999, http://www.wri.org/wr-98-9/malnutri.htm.

149 **As noted in a report** Ibid.

150 **Ironically, the historically** "Australia Damaged by Drought," *New York Times*, October 1, 2003.

150 **The reason involves** "Polar Wind Shift Marks New Weather Worry," Reuters News Service, December 17, 1999.

150 **In Australia, however** "Scientists See Antarctic Vortex as Drought Maker," Reuters News Service, September 23, 2003.

Chapter 7: Thinking Big: Three Beginnings

153 **To me the question of the environment** Hans Blix, quoted in the *Oregonian*, March 14, 2003.

155 **In June 2003** "Europe Adopts Climate Emissions Trading Law," Environmental News Service, July 23, 2003; "A Regional Approach to Global Warming, Romney Joins Group; Environmentalists Laud Effort After U.S. Inaction," *Boston Globe*, July 25, 2003.

155 **A slew of recent scientific findings** "U.S. Greenhouse Gas Emissions Not Offset by Carbon Sinks," Environmental News Service, March 17, 2000; P. Falkowski et al., "The Global Carbon Cycle: A Test of Our Knowledge of Earth as a System," *Science*, October 13, 2000; "Studies Challenge Role of Trees in Curbing Greenhouse Gases," *New York Times*, May 24, 2001; Richard A. Gill et al., "Nonlinear grassland responses to past and future atmospheric CO_2," *Nature* 417, May 16, 2002; "Global Warming Is Changing Tropical Forests," Environmental News Service, August 7, 2002; Christine L. Goodale and Eric A. Davidson, "Carbon Cycle: Uncertain Sinks in the Shrubs," *Nature* 418, August 8, 2002; "Tree Farms Won't Halt Climate Change," *New Scientist*, October 28, 2002; "A Fading Green Hope for Climate," *U.S. News & World Report*, February 10, 2002; "Old Trees Poor Carbon

Sponge? Carbon Stockpiles Question Idea That Forests Will Counteract Global Warming," *Nature News Service*, July 23, 2003; Scott R. Saleska et al., "Carbon in Amazon Forests: Unexpected Seasonal Fluxes and Disturbance-Induced Losses," *Science*, November 28, 2003.

155 **One team of seventeen scientists** P. Falkowski et al., "The Global Carbon Cycle."

155 **Another team of scientists** "A Fading Green Hope for Climate," *U.S. News & World Report*, February 10, 2002.

158 **Under the plan** Aubrey Meyer, http://www.gci .org.uk.

160 **The proponents of** Tom Athanasiou and Paul Baer, *Dead Heat: Global Justice and Global Warming* (New York: Seven Stories Press, 2002).

161 **A second large-scale proposal** Peter Barnes, *Who Owns the Sky?* (Washington, DC: Island Press, 2001).

164 **The most recent large-scale attempt** Author's conversations in 2003 with Michael Shellenberger, Adam Wehrbach, and Bracken Hendricks of the Apollo Project.

167 **Unfortunately, deep-mine burial** "Policy Context of Geologic Carbon Sequestration," Web site of The Union of Concerned Scientists, 2002, http://www.ucsusa.org/global_ environment/global_warming/page.cfm?pageID=529.

Snapshots of the Warming No. 7

171 **In November 2000** "One Thousand Flee as Sea Begins to Swallow Up Pacific Islands," *Independent* (UK), November 29, 2000.

172 **A similar evacuation** "Tuvalu: Global Warming's First Casualty," *Honolulu Star Bulletin*, August 19, 2001.

172 **As a result, the president** "NZ Would Throw Tiny Tuvalu a Lifeline, Says Goff," Reuters News Service, June 21, 2000.

172 **The increasing temperatures** "Tuvalu: Global Warming's First Casualty," *Honolulu Star Bulletin*, August 19, 2001.

172 **President Falcam was** Ibid.

172 **Warming is most pronounced** "Alaska, No Longer So Frigid, Starts to Crack, Burn, and Sag," *New York Times*, June 16, 2002.

173 **At the other end of the world** "Bizarre Weather Ravages Africa Crops, Some See Link to Worldwide Warming Trend," *Washington Post*, January 7, 2003.

Chapter 8: Rx for a Planetary Fever

175 **Our house is burning down** Chirac address to World Summit on Sustainable Development, Johannesburg, South Africa, September 2, 2002.

177 **Dr. Rajendra Pachauri** "IPCC Chief: Global Warming May Nurture Extremism," Reuters News Service, December 9, 2002.

178 **It is an article of faith** Author's conversation with Dr. Morris Miller, former executive director of the World Bank, 1999, and other specialists in energy and development.

178 **Beyond the threat of terrorism** Michael T. Klare, *Resource Wars: The New Landscape of Global Conflict* (New York: Henry Holt, 2001), p. 29.

180 **In the 1980s** "1998 a Disaster for Insurers, Leading Firm Says," Reuters News Service, January 18, 1999, quoting officials of Munich Reinsurance.

181 **As noted earlier** "Climate-Related Perils Could Bankrupt Insurers," Environmental News Service, October 7, 2002; "Global Warming to Cost $300 Billion a Year," Reuters News Service, February 4, 2001; "Climate Change Costs Could Top $300 Billion Annually," Environmental News Service, February 5, 2001; "Climate Change Could Bankrupt Us by 2065," Environmental News Service, November 24, 2000.

182 **The plan was developed** Participants in early discussions of the World Energy Modernization Plan included: Dr. Frank Ackerman, Global Development and Environment Institute, Tufts University, Medford, MA; Dr. Steven Bernow (deceased), vice president, Tellus Institute, Boston; Thomas R. Casten, CEO, Trigen Energy Corporation, White Plains, NY; Dr. Michael Charney, Cambridge, MA; Stephen Cowell, CEO, Conservation Services Group Boston; Dr. Paul Epstein, Associate Director, Center for Health and the Global Environment, Harvard Medical School, Boston; Ross Gelbspan, author, *The Heat Is On*, Brookline, MA; Dr. Jonathan Harris, Global Development and Environment Institute, Tufts University; Ted Halstead, founder, Redefining Progress, Washington, DC; Sivan Kartha, Stockholm Environment Institute, Boston; Dr. David Levy, School of Management, University of Massachusetts, Boston; Dr. William Moomaw, director, International Environmental Research Program, Tufts University;

Dr. Irene Peters, economist, Zurich, Switzerland; Dr. Kilaparti Ramakrishna, Woods Hole Research Organization, Woods Hole, MA; Kelly Sims, Ozone Action, Washington, DC.

183 **Most recently, it was endorsed** Author's conversations with Sir Crispin Tickell, former British ambassador to the UN, Cambridge, MA, 2002.

183 **As Britain's prestigious Institute** "Kyoto Will Not Stop Global Warming," http://www.BBCNews.com, August 8, 2003.

183 **Tony Grayling** Ibid.

183 **Those sentiments were echoed** Ibid.

185 **On the issue of subsidies** Douglas Koplow and Aaron Martin, *Fueling Global Warming: Federal Subsidies to Oil in the United States* (Washington, DC: Industrial Economics, June 1998).

186 **One attractive source of revenue** Mahbub ul Haq, Inge Kaul, and Isbelle Grunberg, eds., *The Tobin Tax: Coping with Financial Volatility* (New York: Oxford University Press, 1996).

187 **If a Tobin Tax proves** Author's conversation with Florentin Krause and Stephen DeCanio, July 11, 2002.

189 **The solution to this dilemma** Amory B. Lovins, "Twenty Hydrogen Myths," Rocky Mountain Institute, July 12, 2003.

197 **We are cutting down forests** Janet M. Abramovitz, *"Taking a Stand: Cultivating a New Relationship with the World's Forests,"* Worldwatch Paper No. 140, Worldwatch Institute, April 1998.

198 **According to a recent study** Ransom A. Myers and Boris Worm, "Rapid Worldwide Depletion of Predatory Fish Communities," *Nature* 423, May 16, 2003.

198 **Each year, we release** Theo Colborn, Dianne Dumanoski, and John Peterson Myers, *Our Stolen Future* (New York: Dutton Books USA, 1996).

Index